DO 0048644 2

AF616125

HARRY L. HELMS

Technical Writer and Editor

HANDBOOK OF PRACTICAL I.C. CIRCUITS

PRENTICE-HALL INC., ENGLEWOOD CLIFFS, NEW JERSEY 07632

Editorial/production supervision
and interior design: Sue Dib
Cover design: Wanda Lubelska Design
Manufacturing buyer: Gordon Osbourne

Printed in the United States of America

10 9 8 7 6 5 4 3 2 1

0-13-380833-5 025

Prentice-Hall International (UK) Limited, *London*
Prentice-Hall of Australia Pty. Limited, *Sydney*
Prentice-Hall Canada Inc., *Toronto*
Prentice-Hall Hispanoamericana, S.A., *Mexico*
Prentice-Hall of India Private Limited, *New Delhi*
Prentice-Hall of Japan, Inc., *Tokyo*
Prentice-Hall of Southeast Asia Pte. Ltd., *Singapore*
Editora Prentice-Hall do Brasil, Ltda., *Rio de Janeiro*

This book is dedicated to Forrest M. Mims III, who has done more than any other person to demystify IC technology for electronics hobbyists and the public at large. Thanks, Forrest!

contents

preface

For those of us who became interested in electronics in the days of vacuum tubes and transistors, today's integrated circuits (ICs) are nothing short of miraculous. Entire circuits that once required dozens of discrete components, together with considerable construction and "debugging" time, are now available in a postage-stamp-size package. Obtaining a desired circuit function, such as amplification or a logic gate, today involves little more than selecting an appropriate IC and adding a few external components (such as capacitors and resistors) together with a voltage source. All the hard work has already been done! ICs are, in effect, little "packages" of amplification, voltage regulation, detection, timing, inverting, and similar useful functions.

The nature of electronic circuit design and construction has also changed. Circuit design is much more of a "cookbook" procedure, requiring far less complex math, voltage and current measurements, attention to parts layout and placement, and so on, than in previous years. Designing with IC devices is largely a matter of "chaining" appropriate ICs and the functions they perform together to create working electronic devices. Moreover, the availability of inexpensive and versatile "breadboard" design tools makes experimentation with ICs and circuit prototyping merely a matter of inserting ICs and other components into holes and connecting them with "plug-in" wires. Circuit construction now takes minutes rather than hours of careful soldering. Your author really feels old when he contemplates that many younger electronics enthusiasts have never known the "joy" of trying to solder parts mounted on a so-called "perfboard," or the smell of a fried transistor that was inadequately heat-sinked during soldering. (If you are one of those younger people, take it from your author—you haven't missed a thing!)

This book is a collection of IC "recipes." The emphasis here is not on designing IC applications circuits "from scratch," but rather on working, debugged circuits that are ready to be used. Over the years, many professional electronics engineers, as well as a surprising number of electronics hobbyists and tinkerers, have developed numerous de facto "standard" or "building-block" circuits around popular ICs. It is certainly possible to develop your own original circuits to accomplish the same functions, but why reinvent the wheel? The circuits in this book are ready to go or can easily be modified for your specific needs. The result is that your time is spent more productively on original, creative work of your own rather than chasing down the solution to a problem that somebody else has already solved.

In addition to the IC circuits themselves, this book discusses the basic theory behind all ICs and traces their evolution. There is also information on the various types and "families" of IC devices available. You will find descriptions of how popular ICs operate and details on circuit prototyping and testing. Reference data on the functions contained on popular chips, together with their "pin-out" connections, have been included.

The development of ICs did more than just revolutionize electronics; it also "democratized" electronics. A high school student has the same access to and ability to experiment with some of the technology as that available to a Ph.D. in electrical engineering. It is now possible for inventive and creative hobbyists to come up with devices and inventions that are more innovative and significant than some of those developed by billion-dollar firms. That statement may seem overblown, but consider the personal computer revolution. It did not begin in the research and development departments of major companies such as IBM or GE, but instead, in the garages and basements of such people as Steve Wozniak and Ed Roberts. Although they did not create the basic technology—the microprocessor—they saw and exploited a potential that the professionals overlooked.

Hopefully, this book will stimulate you into devising some interesting and creative applications of IC technology.

Harry L. Helms

chapter 1

INTRODUCTION TO INTEGRATED CIRCUITS

One of the seminal events in contemporary electronics may well have taken place in 1958, when Jack Kilby was ruled ineligible for summer vacation from his job as an engineer at Texas Instruments. A new employee, Kilby had not accumulated enough service time with Texas Instruments (or "TI" as it is known in the electronics industry) to qualify for a paid vacation during the company's scheduled annual vacation period in 1958. While most of his co-workers were taking it easy or driving the family to the beach, Kilby was busy working on a pet project to fabricate a complete working circuit on a single section of germanium. During that summer of 1958, he succeeded in "integrating" the complete circuit for a phase-shift oscillator onto a germanium slice, and by September of that year had a functioning prototype completed. TI filed for a patent in 1959, and by 1961 had integrated circuits (ICs) in production in commercial quantities. Kilby was not the only engineer hard at work on circuit integration around that period, however. Robert Noyce, then with Fairchild Semiconductor, developed an improved process for circuit integration and also filed for a patent in 1959. Noyce later became one of the cofounders of Intel, a major technical innovator in integrated semiconductors.

The term *integrated circuit* came to describe the devices pioneered by Kilby and Noyce. Early ICs were technical marvels because they contained the functional equivalent of two or three transistors on a single chip of silicon. Some current ICs, however, contain the equivalent of several hundreds of thousands of transistors on a single chip, and it seems likely that ICs containing the equivalent of over 1 million transistors on one slice of silicon will soon be available. (One semiconductor industry leader has remarked that although such a feat is within the realm of technical feasibility, he is not sure that engineers would know what to do with such a device.)

IC FABRICATION AND "PACKAGING"

The basic idea behind an IC is simple. The major "parts" of any IC are those that make up almost any circuit, such as transistors, resistors, capacitors, and diodes. In fact, most data sheets for ICs give what is known as an *equivalent circuit*. This is a representation of the circuit "on" the IC as it would be constructed from discrete components. This is the circuit formed on the slice of silicon making up the IC. Moreover, the circuit formed on an IC may actually be several identical working circuits (such as logic "gates" or timers) formed on the same slice of silicon; such circuits normally function independently of each other. It is possible to use all or just one of these circuits on an IC. Thus you may see a reference to something like a "1/2 556" or "1/4 7400"; this means that only a portion of the circuit functions available on an IC are being used as part of a larger circuit design.

IC transistors and diodes can be formed from junctions of P-type and N-type material, as is done with discrete transistors and diodes. IC transistors may be bipolar or implemented using another technology, such as metal-oxide semiconductor (MOS), field-effect transistors (FETs), or MOSFETs; most bipolar transistors are NPN. Resistors are formed on ICs from a very small section of P-type material, while capacitors can be formed across the depletion zone of a PN junction that is reverse biased. Inductors are seldom found on ICs, since they are difficult to integrate successfully; the Q value of such inductors is usually quite poor. IC designers use a variety of techniques to avoid inductors and to stimulate inductive effects where necessary.

ICs are formed on a ''wafer'' of P-type silicon known as a *substrate.* Over 100 ICs can be formed on a single wafer; the finished wafer is cut into sections to produce the individual ''chips.'' The wafer is coated with a thin layer of silicon dioxide, a glasslike insulating material. Additional thin layers of silicon doped with P-type and N-type material are placed on the wafer through a process known as *epitaxy.* The circuit design is ''cut'' into the wafer by a photomasking and etching process which produces the various individual circuit elements, such as transistors and resistors. However, these individual elements are still electrically isolated from each other. The elements are connected by thin slivers of aluminum, which act as wiring for the IC. These aluminum slivers are connected to the pins of the IC package. Figure 1–1 shows how an NPN transistor is connected to a resistor on an IC.

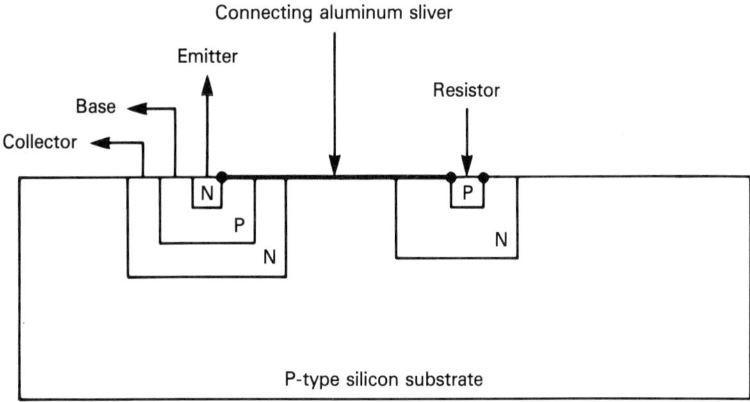

Figure 1–1 NPN transistor and resistor integrated on substrate of P-type silicon.

ICs are often described as being *monolithic* or *hybrid.* A monolithic IC is a complete circuit performing a specific function integrated onto a single chip of silicon. A hybrid IC is formed from two or more silicon sections, performing different functions (whether active or passive), which are connected together to form the final working circuit.

ICs may also be described as *thick film* or *thin film.* A thick-film IC is one in which resistors and capacitors are formed on the substrate but transistors are added as discrete elements atop the substrate. A thin-film IC is one in which the substrate is an insulating material such as glass or ceramic; all components are formed atop this substrate. The advantage of this approach is that the electrical isolation of the individual elements is superior to that of ICs using a silicon substrate.

All ICs, once separated from the wafer, are encased in a protective package. There are three major types of packaging for ICs. One package is a small metal "can" which looks like the housing for a discrete transistor, but which has more leads. This package, known as the *TO-5,* is a rugged housing. Most TO-5-packaged ICs have 8, 10, or 12 leads and an identifying tab on one side. This tab usually indicates the *last* pin number; the first pin immediately to the left of the tab is pin 1 of the IC. Pin numbers run counterclockwise until the last number is reached. Figure 1–2 shows the usual pin arrangement for a TO-5 case.

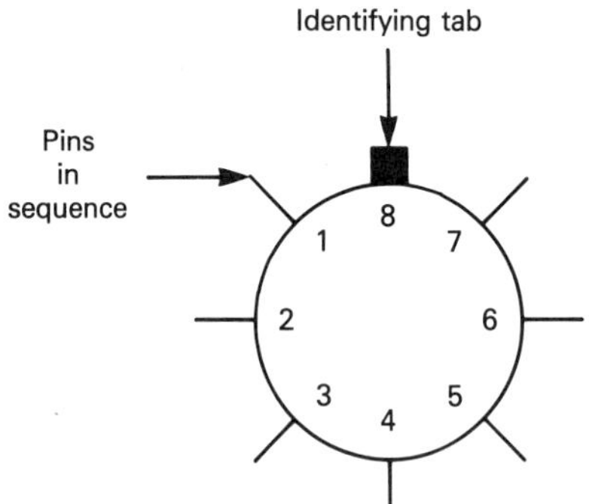

Figure 1–2 Identifying tab and pin numbering sequence for a TO-5 IC package.

Another IC packaging is the TO-81, better known as the *flat pack.* The flat pack is made of ceramic or plastic, with the ceramic pack offering a better seal against the surrounding environment. A matching set of flat pins extend outward from the package. Flat-pack ICs are marked with a dot to indicate the location of pin 1, and pin numbering proceeds in the order indicated in Figure 1–3. Most flat-pack ICs have 8, 14, 16, 24, or 28 pins.

However, ICs are most commonly found in rectangular black plastic or ceramic cases with a row of pins running along two sides of the package; this is known as *dual-in-line package* (DIP). Its more formal designation is "the TO-116 package." Figure 1–4 shows a diagram of a typical DIP.

If you plan to do any work whatsoever with ICs, the ability to "read" a DIP is essential. One end of the DIP is normally marked with an indenta-

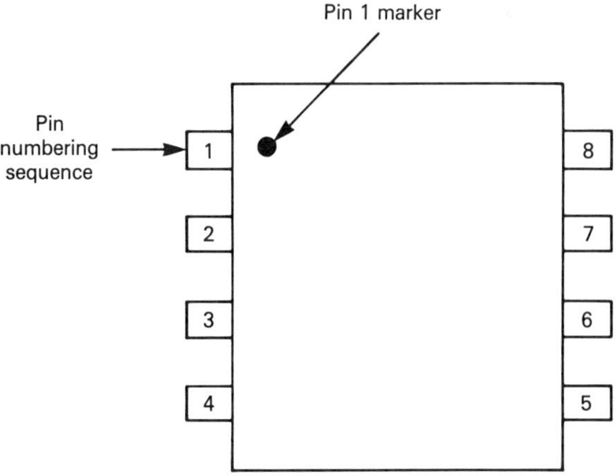

Figure 1-3 Pin 1 identifying mark and pin numbering sequence for a TO-81 IC package.

tion of some sort (usually square or semicircular). This indicates which end of the DIP is, as it might be described, "up." The pin in the uppermost left corner from this indentation is pin 1 of the IC. Pin numbering proceeds "down" the left side of the IC and then continues with the uppermost pin

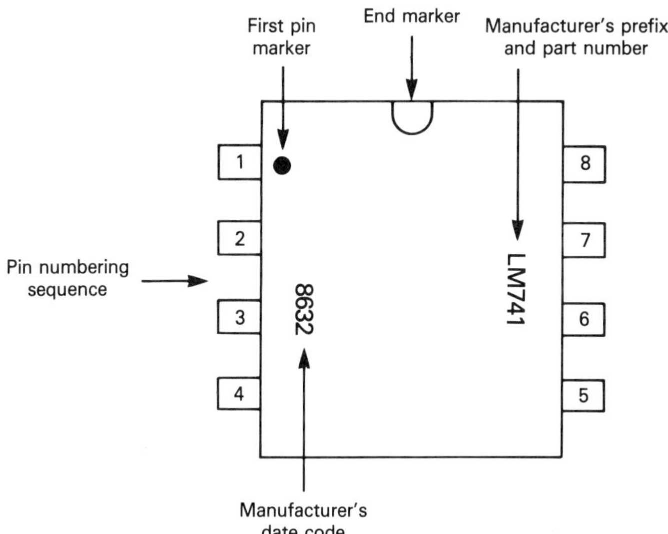

Figure 1-4 TO-116 dual-in-line package (DIP) IC packaging.

TABLE 1-1 IC Manufacturer Prefixes

Prefix	Manufacturer
AD	Analog Devices
Am	Advanced Microdevices
CA, CD	RCA
DM	National Semiconductor
FE	AND, Inc.
FSS	Ferranti
GIC, GP	General Instrument
H	Harris Semiconductor
HA	Hitachi
I	Intel
ICL, ICM	Intersil
IDT	Integrated Device Technology
L, LD	Siliconix
LF, LH, LM	National Semiconductor
LT	Linear Technology
M	Mitsubishi
MB	Fujitsu
MC	Motorola
MCS	MOS Technology
MK	Mostek
MM	Monolithic Memories (sometimes used by Motorola as well)
MN	Micro Networks
MPV	TRW
MT	Mitel (also used by Alpha Industries)
N, NE	Signetics
PM	Precision Monolithics
S	American Microsystems
SE	Signetics
SG	Silicon General
SN	Texas Instruments
SP	Plessey
TDC	TRW
TLO, TMS	Texas Instruments
TSC	Teledyne Semiconductor
WD	Western Digital
XR	Exar, Inc.
μA	Fairchild
μPB, μPD	NEC

to the right of the indentation. Some ICs will have a dot or other small marker adjacent to pin 1, but not always.

Typically, the largest lettering on the IC will be for the device's part number; often, this will be preceded by the manufacturer's prefix. Table 1–1 includes a list of the more common prefixes. Some of these will quickly become second nature to you, and you will automatically associate "MC" with Motorola and "SN" with Texas Instruments. (A few companies also include their corporate logos on the ICs they manufacture.) You will also discover that some popular devices are manufactured by several different companies. In such cases, it is common to refer to the device by its part number alone, as in "741" or "7400." Such devices from different manufacturers are functionally equivalent to each other.

Some manufacturers include data codes on ICs to indicate when they were produced. Usually, the year of manufacture will be indicated by the last two digits of the year, as in "87." There will also usually be one or two digits to indicate when in that particular year the IC was manufactured. Depending on the manufacturer, this code could represent the month or week the IC was produced or the production batch of which the IC was a part. In most cases the date code will be of no particular significance to users, although it can be useful if a particular production run turns out to have an abnormally high percentage of defective products.

IC TYPES AND FAMILIES

Most ICs fall into two major categories. *Linear* ICs respond to or amplify signals or voltages which vary in a continuous form, similar to an alternating current (ac) waveform. On the other hand, *digital* ICs process signals composed of just two voltage levels, known as "high" (above a specified level) and "low" (below a specified level). Some ICs combine both linear and analog functions in a single package.

One classification method for ICs is the *scale of integration,* which refers to the number of transistors or logic gates contained on a chip. *Small-scale-integration* (SSI) devices generally have 10 or fewer transistors or gates on the chip. *Medium-scale-integration* (MSI) devices have 10 to 100 transistors or gates, while *large-scale-integration* devices have over 100 gates or transistors.

The most widely used and versatile linear IC is the *operational amplifier,* fondly known as the "op amp." These ICs amplify the *difference* beween two signals applied to its two inputs. One of these inputs is known as the *inverting* input (labeled with a −) and the other is the *noninverting* input (labeled with a +). Figure 1–5 shows the schematic symbol for an op amp.

Whenever an op amp is used, signals must be applied to both inputs, or if a signal is applied to just one input, the unused input must be connected to ground or to a constant-voltage source. (If the unused input is allowed to "float," the circuit will operate improperly or not at all.) Op amps can be configured for a wide variety of applications, including inverting and noninverting amplifiers, comparators, timers, voltage regulators and references, mixers, and even analog computers.

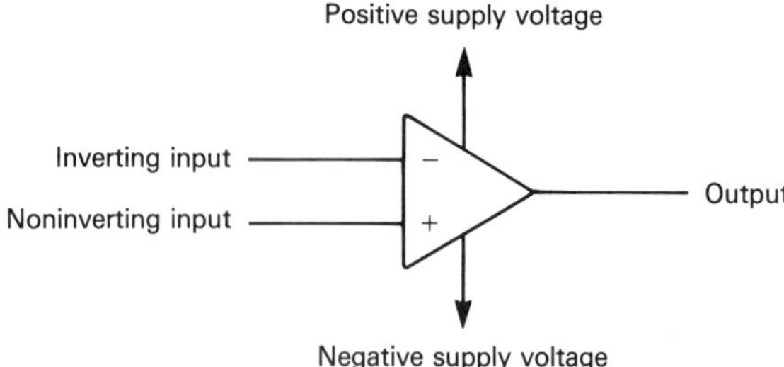

Figure 1-5 "Op amp" schematic symbol with input "polarities."

Op amps whose equivalent circuits are built from bipolar or MOSFET transistors are available, and these are known (appropriately enough) as "bipolar op amps" or "MOSFET op amps." Some bipolar op amps with MOSFET inputs for high input impedance are available. Up to four op amps may be integrated onto a single chip and housed in one package.

Several of the circuits that can be built using op amps are also available as ICs themselves, including timers, voltage regulators, and function generators. Other special-purpose linear ICs have been developed for specific purposes, such as radio, television, and telephones (including ringers and electronic "bells"). One area of rapid growth in linear devices has been *phase-locked-loop* (PLL) ICs. A phase-locked loop circuit is an oscillator whose output follows a reference signal; the output and reference signals are continuously compared and the output signal is "corrected" whenever it does not faithfully "track" the reference signal. PLL ICs are now widely used in communications equipment.

Digital ICs come in different logic *families.* These various families indicate how equivalent circuits using discrete components for logic circuits implemented on ICs are built. In fact, many of the early logic ICs were virtually identical (except for being integrated on a single silicon segment) to those logic circuits engineers had been using for years. Over a dozen different significant families have been created, but in this chapter we examine only the major ones.

The circuits integrated on digital ICs are known as *logic gates.* A logic gate takes two or more digital input signals (i.e., signals that are high or low as defined above) and produces a single high or low output signal depending on the combination of high and/or low input signals present. The exception is the NOT gate, which has only one input and one output. The basic logic gates are known as AND, OR, NOR, NAND, and NOT gates (NOT gates are also known as *inverters*). The basic functions of these are as follows:

AND. Produces a high output only if *all* inputs are high; otherwise, the output will be low.

NAND. Produces a low output only if all inputs are high; if any or all inputs are low, the output is high.

NOR. Produces a high output only if *all* inputs are low; otherwise, the output is low.

NOT. Produces an output that is the opposite of its input; a high input produces a low output and a low input causes a high output.

OR. Produces a low output only if *all* inputs are low; otherwise, the output will be high.

Figure 1-6 shows the symbols for each of these gates.

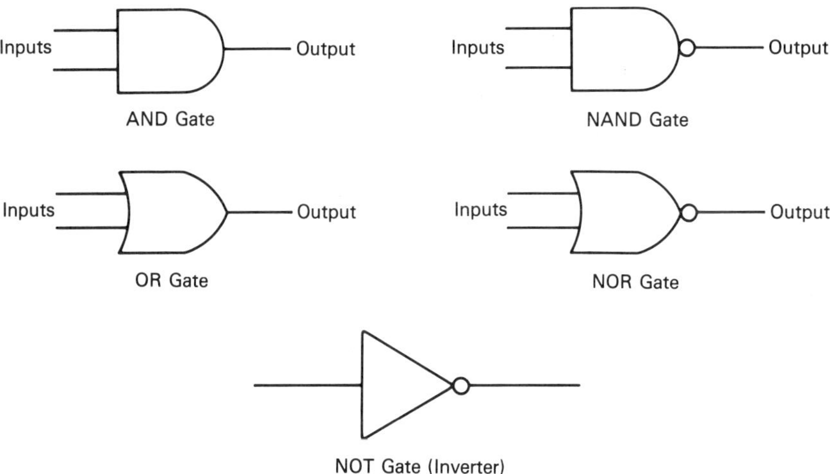

Figure 1-6 Schematic symbols for major logic gate types.

Each of the basic gates functions the same way in each device family. Further, these five basic logic functions are usually combined to make up *sequential* circuits, which are sequences and combinations of the various basic functions to perform desired tasks and create new gates, such as the XOR, which produces an output if all inputs are the same (all high or all low) but not if any are different. Sequential circuits can be put together using separate logic chips, or complete sequential circuits are available on ICs. Many of these sequential circuits are highly complex, but all are still little more than combinations of just five logic functions. In fact, it is possible to build *any* logic circuit using only NAND gates.

Although logic functions are the same between various families, there are numerous differences in how well each family performs when implementing those functions. Here are some of the factors considered when assessing the performance of a device family:

Power requirements. Some IC families can operate over a wide range of supply voltages, whereas others must be powered from a supply voltage held to a fixed, constant level [usually, 5 volts (V)]; if this is exceeded, the IC may be destroyed.

Power consumption. How much current does the device consume?

Fan-in and fan-out. Fan-in refers to the maximum number of inputs a device can have, and fan-out refers to the number of other gates for which a particular gate's output can serve as the input signal (i.e., to *drive* other devices).

Propagation delay. Whenever a change is made to the inputs of a logic gate, there is always a delay before the corresponding change in the output takes place. This factor expresses how long this delay is. Another term for propagation delay is "speed." This might at first seem like a meaningless measure, since even the "slowest" logic devices have propagation delays measured in nanoseconds. However, it is very important in sequential circuits.

Noise immunity. All inputs to a device are plagued by "noise." This means that each input will be subject to transient pulses, voltage variations, and similar fluctuations which can trigger a false high or low input to a gate. This indicates how well the device can ignore signals that do not meet the intended levels for high and low signals.

There are other factors used in evaluating families and devices, but these are usually the most significant.

The term "family" refers to how various IC devices implement logic functions using the integrated equivalents of logic gates constructed on discrete components. How such equivalent circuits work and their schematic diagrams are not of interest to us (there is, in any case, nothing one can do to alter how they work) and will not be discussed here. If you are interested in understanding how the equivalent circuits work, a college-level engineering text on IC design should be consulted. For now, rest assured that the following types of equivalent circuits have indeed been integrated on silicon to produce working logic gates.

The first family to go into commercial production was *resistor-transistor logic* (RTL), in which the logic gate was composed strictly of transistors and resistors. RTL devices had several problems, such as poor noise immunity, low fan-in and fan-out capabilities, and modest speed. RTL required a supply voltage of +3.0 to 3.6 V. Today, RTL ICs are rarely encountered except in older electronics designs, and RTL ICs are seldom seen even on the surplus IC market.

The next logic family to go into commercial production was *diode-transistor logic* (DTL). Despite its name, a DTL device used resistors as well as transistors and diodes. (Perhaps a more accurate name would have been "diode-resistor-transistor logic.") Compared to RTL, DTL offered greater fan-in and fan-out capabilities, improved noise immunity, and higher speed. Power supply requirements were 5 V, with a permitted variation from this figure of less than 10%.

A variation of basic DTL was *high threshold logic* (HTL). In HTL, higher input signals are required than with conventional DTL; this gave the IC greater noise immunity than ordinary DTL. This enhanced noise immunity was obtained at the cost of lower operating speed, however. Moreover, HTL required a 15-V power supply and DTL and HTL devices could not be "mixed" in a circuit (i.e., the output of a DTL IC could serve as the input for an HTL chip, and vice versa).

Like RTL, DTL logic is seldom encountered today and HTL is seldom used. RTL, DTL, and HTL ICs were superseded by *transistor-transistor logic* (TTL) and *complementary MOS* (CMOS) devices. In fact, it is likely that you may never need to use another type of logic IC other than TTL and CMOS.

TTL ICs were the next major family developed after DTL and share some similarities with DTL. Both, for example, require a constant 5-V power source and they are "compatible" with each other; the output of a DTL IC can serve as the input for a TTL device, and vice versa. It is significantly faster than DTL and has modestly larger fan-in and fan-out capabilities. It also has good noise immunity, but is sometimes more sensitive than DTL to short-duration pulses because of its greater speed.

The popularity of TTL resulted in numerous ICs being produced using that technology, and TTL is still widely used today; moreover, TTL devices are relatively inexpensive. It will probably remain a popular device family for several more years. However, TTL has some drawbacks. The major one is current consumption. TTL ICs are not well suited for applications where the primary power source will be a battery. In addition, the operation of TTL is such that when a gate switches between high and low levels, a low impedance can be "shown" to the power supply, producing a brief, heavy demand on the power supply, a momentary "glitch," or other transient in the power supply. This can show up as "noise" and affect other devices powered by the same supply.

One variation of "standard" TTL that has achieved wide use is *low-power Schottky TTL* (LSTTL). In such devices, a different type of diode, known as a "Schottky barrier" device, is integrated on the chip. This produces an IC with the same speed as standard TTL but much lower current consumption. Indeed, in most applications LSTTL has largely replaced standard TTL. Whether an IC is TTL or LSTTL is indicated in its number. For example, a popular TTL IC is the 7400, which has four NAND gates in a single package. The LSTTL equivalent is identified as 74L500. All TTL and LSTTL ICs have part numbers beginning with the digits "74."

The biggest advance in IC technology, however, was the introduction of CMOS devices. These devices use the least current of any digital family. For example, a gate on a typical TTL device may require over 3 milliamperes (mA) of current, whereas the same gate on the equivalent CMOS device usually draws less than 0.1 mA. In addition, CMOS devices can operate over a wide power supply range, typically 3 to 18 V, with no adverse effects on its performance. Noise immunity is better than that for TTL, and CMOS produces less noise of its own in operation than does CMOS. Today there are CMOS equivalents of virtually every TTL IC.

However, CMOS is not the perfect family. All CMOS devices are prone to damage by static discharge. This means that such devices must be handled, installed, and stored with care to prevent them from being "zapped" by static electricity. (If you do much work at all with CMOS, however, you might as well reconcile yourself to the loss of an IC from time to time, despite your best efforts.) CMOS devices are also still more expensive, on average, than comparable TTL ICs. More serious is the fact that "plain vanilla" CMOS is signficantly slower than TTL. More recent versions of CMOS, known as *high-speed* CMOS, offer operating speeds as good as those of TTL. With this new variation, CMOS seems destined to replace TTL just as TTL replaced the DTL and RTL families.

CMOS equivalents of popular TTL devices are available, and these are identified by the letter C in the middle of the TTL device number. The CMOS equivalent of the 7400 device used in the previous example would be identified as 74C00. Original CMOS devices have part numbers beginning with the number "4." However, the equivalent TTL and CMOS part numbers do not mean that the devices are equivalent. The CMOS equivalent of a TTL device will always follow the TTL numbering system but with the letter C to denote that it belongs to the CMOS family. New devices belonging to the high-speed CMOS group are designated by the letters "H" or "HC" in the middle of the part number; the high-speed CMOS equivalent of the 7400 TTL IC would be designated as 74HC00.

CMOS and TTL devices are not compatible, however, even when equivalent devices (such as the 7400 and 74C00) are involved. When they are used together in a design, appropriate interface circuits must be used. Such circuits are given in following chapters.

There are additional IC families used in special applications. Two are variations of TTL, known as *emitter-coupled logic* (ECL) and *integrated injection logic* (IIL or I^2L). ECL is the fastest form of digital logic in wide use today, with speed far in excess of conventional TTL and CMOS (including high-speed CMOS). Unfortunately, ECL has a very low noise immunity and it requires two different voltage sources. I^2L technology allows circuits to be integrated without resistor-equivalent elements; this reduces the heat produced from the device and permits a "denser" IC (in terms of the numbers of transistors). An I^2L IC typically has over 100 times as many transistors as a TTL chip of the same size. Such density comes at a much greater cost than TTL and CMOS, though.

Other MOS technology ICs include *P-channel MOS* (PMOS) and *N-channel MOS* (NMOS). These are often used in more complex ICs such as microprocessors and memory devices, and are beyond the scope of this book. They are slower than TTL devices and, in fact, they are few equivalents to TTL ICs in the PMOS and NMOS families. Like ECL, these often require two different voltage sources.

Some ICs combine both analog and digital functions in one package. Among the most useful devices in this category are the *analog-to-digital converter* (ADC) and *digital-to-analog converter* (DAC). These devices, as their names imply, take a signal in one form (analog or digital) and convert it to the opposite form. Other ICs involving analog and digital functions in the same package include music and voice synthesis devices.

CONSTRUCTING IC CIRCUITS

IC circuit prototyping can be done in minutes through the use of *solderless modular sockets,* also known as "breadboards." These units allow ICs and other components to be inserted and connected together through conducting terminal construction strips under the socket's surface or by jumper wires. Figure 1-7 shows a typical solderless modular socket, and Figure 1-8 shows the pattern of the conducting common connection strips under the surface. Since no soldering or other permanent connection is made to each part, all components can be reused as often as necessary and changes to a circuit involve nothing more than plugging in the new parts and perhaps a couple of additional jumper wires.

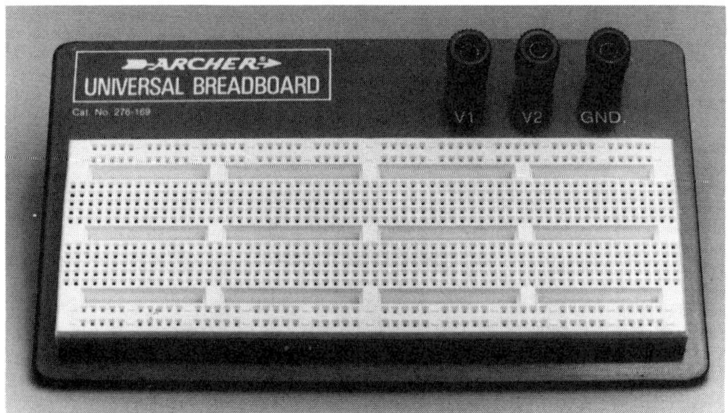

Figure 1-7 Solderless modular breadboard available from Radio Shack. Note connection points for two different voltages.

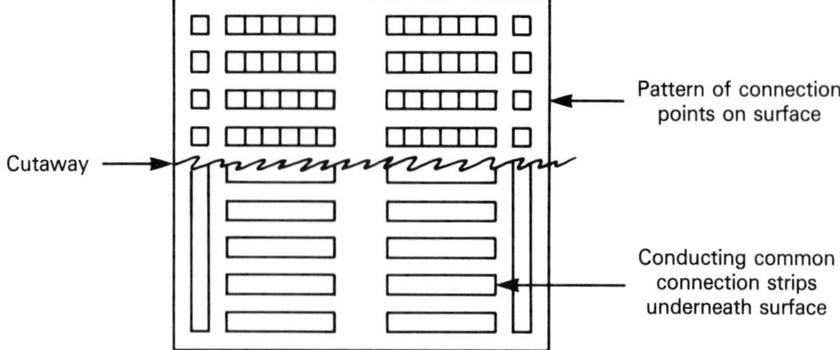

Figure 1–8 Cutaway view of a solderless breadboard showing the pattern of conducting strips under the surface.

Prototyping breadboards are available in a variety of sizes; these are usually measured in terms of the number of *connection* or *tie points* available on the socket. Some of these come with binding posts for connecting a power supply; a few deluxe breadboards even come with power supplies built in (typically for 5 V), together with supports for additional components, such as potentiometers, LEDs, and meters.

Breadboards are essential for circuit design and experimentation. Many hobbyists, in fact, use nothing else when working with ICs. However, breadboards are totally unsuitable for even semipermanent versions of circuits, as parts and connecting wires are easily knocked loose or even knocked completely out of the connection points. Other methods must be used to produce permanent versions of "breadboarded" circuits.

One method, known as *perfboard,* utilizes a section of phenolic board through which numerous, closely spaced holes have been drilled. Part leads are inserted through the holes and are either twisted together or connected by "jumper" wires before soldering; all connections and soldering are normally done on only one side of the perfboard. Although perfboard provides an inexpensive construction technique, it is not suitable for circuits that could be subjected to stresses such as vibration, and components can be damaged by twisting their leads together.

Printed-circuit (PC) boards are also used in constructing IC circuits. Like perfboards, components are mounted on one side of the board by inserting them through holes on the board. They are connected on the opposite side by soldering the part leads to copper foil strips that trace the circuit path. PC boards produce circuits which are rugged and, once the PC board has been designed, can be assembled easily.

Since perfboards and PC boards require soldering, reusing parts is difficult. A relatively new technique which produces rugged finished circuits and allows parts to be reused is *wire wrapping.* A "wire-wrap" card is covered with IC sockets with short pins protruding from the underside of the wire-wrap card. ICs can be inserted directly into the sockets, while discrete components are first mounted on adapters or "headers" which can be plugged into the sockets. The various components are connected by conducting wires wrapped around the pins attached to each socket connection. The wires are attached to each pin by a wire-wrapping tool; these can be either manual or automatic types. The reliability and strength of a wire-wrapped connection is often equal to that of a soldered connection, yet the chances of damaging an IC or other part are greatly reduced. Changes can be made easily and parts may be reused. Wire wrapping is the quickest method of constructing permanent circuits and is the preferred method for building a few custom circuit designs; if several are needed, PC boards are normally a more cost-effective approach (particularly if automatic soldering equipment is used).

The best construction method depends on your needs. For most experimenters and hobbyists, a solderless IC breadboard will be the main working tool; if a more permanent form is desired, wire wrapping is preferable to perfboard because it allows the components to be reused. PC boards usually make sense only in high-volume-production applications.

If you do use a soldering-based construction technique (such as perfboard or PC boards), it is wise to use IC sockets rather than soldering directly to the IC itself. Soldering directly to ICs requires a finer "touch" than that needed in soldering to transistors or other discrete devices, and many ICs can easily be damaged by the heat used in soldering. A supply of these sockets can be cheap insurance against inoperative ICs if you are not skilled in IC soldering.

POWERING IC CIRCUITS

The exact power requirements for a given IC can be ascertained by consulting the *data sheet* for the device in question. The data sheet produced by the manufacturer of the IC gives such essential data as pin connections, internal diagrams, operating parameters, suggested uses, applications circuits, and related information. Data sheets can be obtained upon request from IC manufacturers or their sales representatives; several annual publications also contain data sheets or condensed summaries of their information. Fortunately, however, most ICs fall into certain categories for their power requirements, and it is usually not necessary to consult a data sheet or other reference. The necessary supply voltages for all ICs mentioned in this book will be given; all such voltages will be direct current (dc).

In most IC applications, supply circuit is less crucial than supply voltage. As mentioned previously, all TTL logic ICs require a fixed, stable 5-V power source. Since most CMOS ICs can operate anywhere in the range 3 to 18 V, 5 V has been unofficially adopted as the "standard" digital or logic

IC supply voltage. The easiest way to obtain this voltage is from a 6-V battery source (such as 4 1.5-V cells in series) with appropriate extra components to "drop" the voltage to approximately the desired 5-V level. Figure 1-9 shows a circuit that can produce approximately 5.1 V from a 6-V battery source.

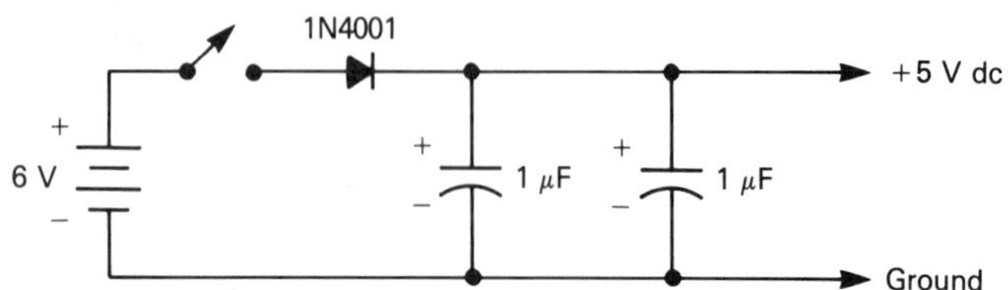

Figure 1-9 Battery power supply providing 5 V for digital ICs.

While the circuit in Figure 1-9 is simple, most TTL devices consume much current and will quickly exhaust any battery supply. An ac line-operated power supply is much more economical in the long run. Figure 1-10 shows a simple ac power supply using a 7805 voltge regulator IC for a constant 5-V output at a little more than 1 A of current. The power transformer can be any type having a 110-V primary with a 8.5-V secondary. The bridge rectifier is made of four 1N4002 rectifier diodes.

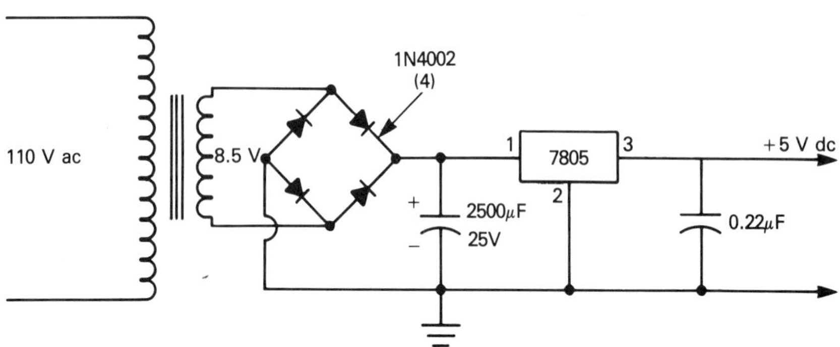

Figure 1-10 110-V ac power supply delivering 5 V for digital ICs.

The situation is not quite as simple with linear devices. Most linear devices, similar to CMOS digital ICs, can operate over a wide range of supply voltages. Not all, however, can operate properly at 5 V. Since design of complex circuits is simplified if all ICs, linear as well as digital, operate at the same voltage, a "movement" has been under way to produce linear ICs that can be powered by 5 V. Among the more popular linear IC devices that operate from 5 V are the 555 and 556 timers, the LM308 and LM312 op amps, the LM311 comparator, and the LM331 voltage-to-frequency converter.

Despite this, many widely used linear ICs, including numerous popular op amps, require dual polarity (+ as well as −) supply voltages. The de facto standard for these dual-polarity voltages has been 15 V, although most such ICs will work just as well from dual-polarity 9-V sources.

The simplest method of obtaining a dual-polarity 9 V power source is shown in Figure 1–11. This is nothing more than two 9-V batteries with their opposite-polarity terminals connected to a common ground point. Leads from this power supply to the op amp(s) should be as short as possible, and less than 6 in. (15 cm) in any case.

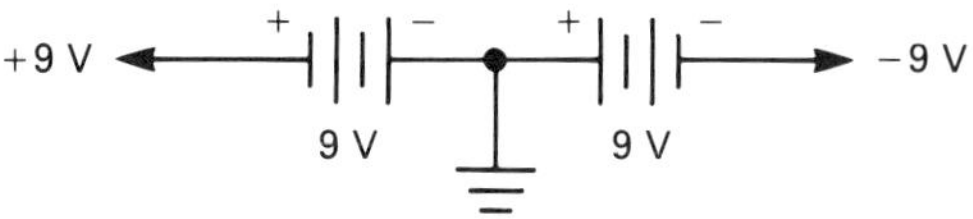

Figure 1–11 Dual-polarity 9-V power supply using two 9-V batteries.

As was the case with digital IC power supplies, a power supply using ac line voltage will save money in the long run and be more convenient (you will not be faced with a dead battery when you want to test a new circuit). Figure 1–12 shows an ac line-operated power supply which provides dual-polarity 9-V sources. Each terminal supplies approximately 60 to 70 mA of current.

If both linear and digital ICs are used in the same design, one often needs a 5-V source for the digital devices and a dual-polarity source at 9 V to 15 V for the digital devices. Figure 1–13 shows a power supply circuit that provides a constant 5-V output for digital ICs while providing adjustable dual-polarity outputs from 1.2 to 17 V. Each dual-polarity output can deliver in excess of 1 A of current.

Numerous power supplies are available commercially to provide the necessary voltages for ICs circuits. Some of the more expensive solderless modular breadboards incorporate digital and linear device power supplies, as do some "IC trainer" devices specifically designed for IC instruction and experimentation.

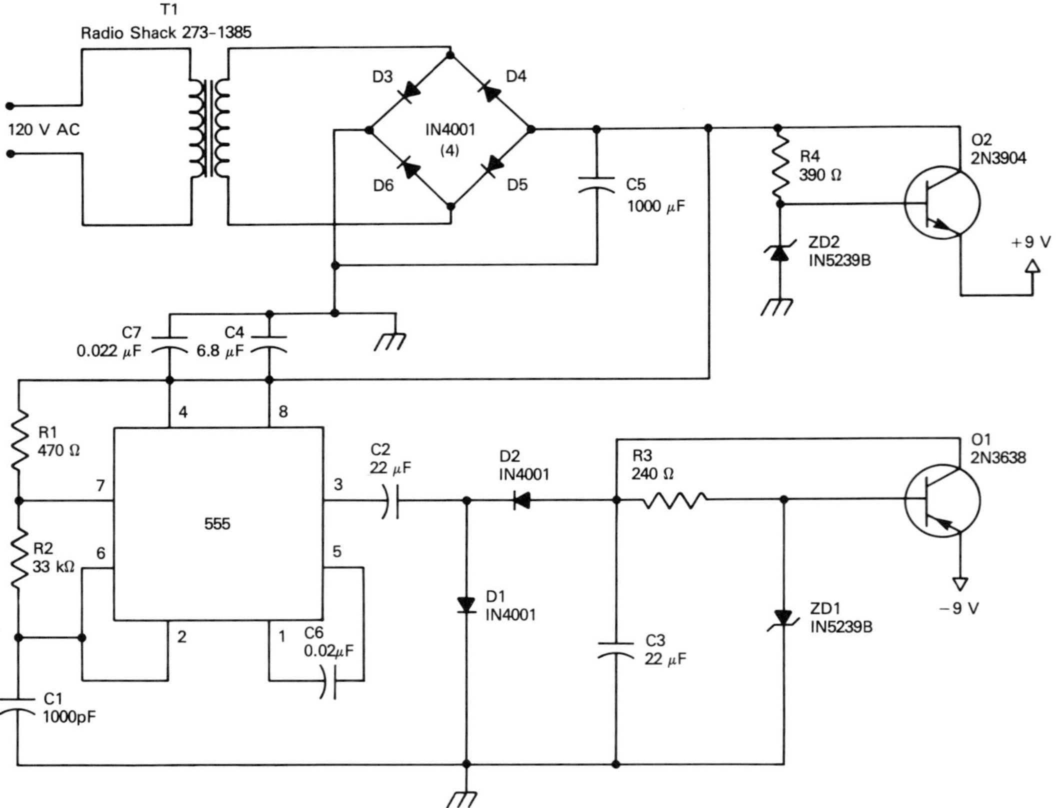

Figure 1–12 Ac line-operated power supply delivering dual-polarity 9-V outputs (courtesy of *73 Magazine*, copyright 1980).

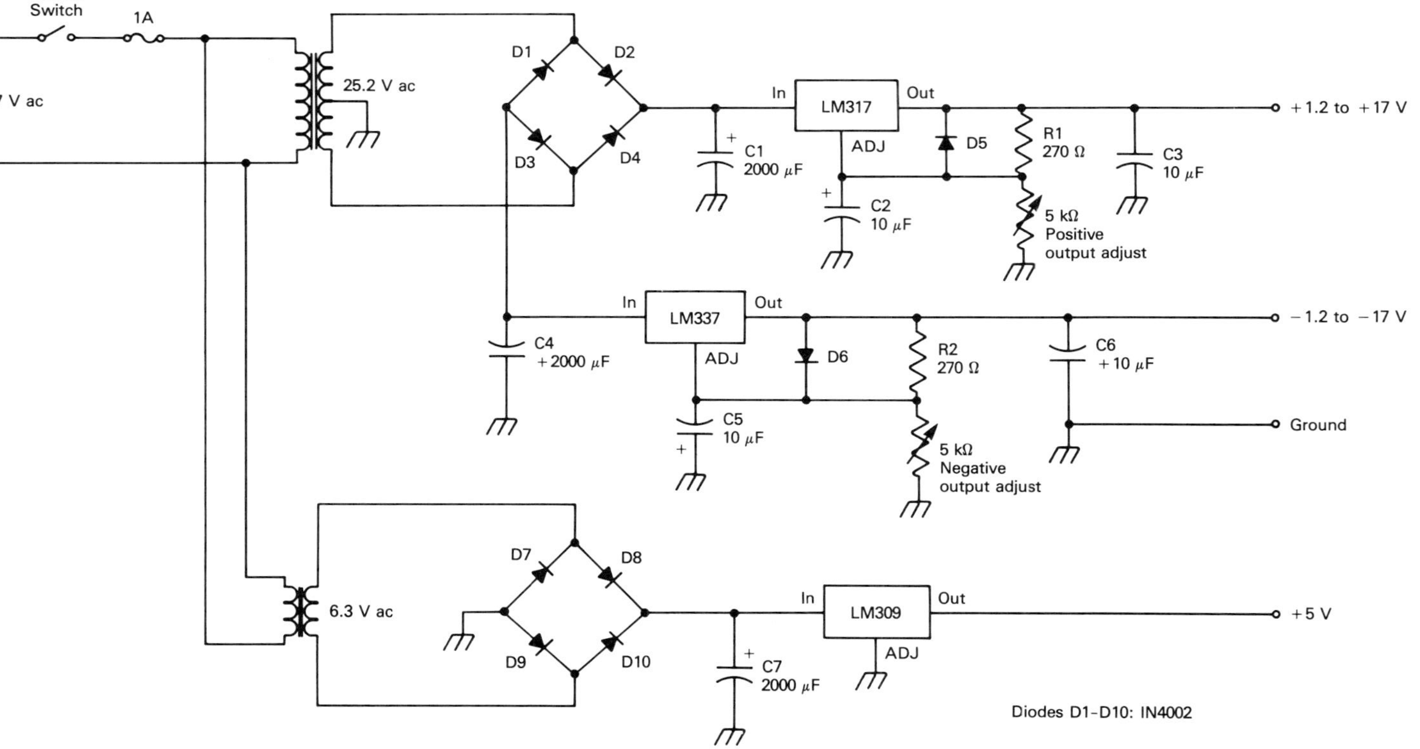

Figure 1–13 Ac line-operated power supply delivering dual-polarity voltages from 1.2 to 17 V for linear devices and constant 5 V for digital ICs (courtesy of *73 Magazine*, copyright 1981).

chapter 2

OPERATIONAL AMPLIFIER CIRCUITS

The ''op amp'' may well be the most versatile type of IC, linear or digital, available today. Op amps show up as the major element in such circuits as amplifiers, oscillators, filters, comparators, function generators, integrators, differentiators, and other analog designs. They can also be configured as various logic gates. Indeed, it is difficult to imagine what contemporary electronics would be like without op amps.

To get the most out of op amps, you must understand some of the basic theory underlying all op amp ICs. Fortunately, the theory and supporting math in this chapter are not too difficult (in fact, this material is about as ''theoretical'' as we will get). Once mastered, the essential concepts are applicable to virtually all op amps currently available.

OP AMP HISTORY AND THEORY

The first monolithic IC op amp, the 709 type, was designed in 1965 by Bob Widlar, then of Fairchild Camera and Instrument Corp. The most popular op amp in use today, the 741, was designed in 1967 by Dave Fullager, then also at Fairchild. Despite the age of the design, the 741 remains the ''standard'' op amp, and many subsequent op amps are basically variations on and enhancements of the 741. The development of the operational amplifier concept goes back over four decades, however. It is generally agreed that the concept of the operational amplifier was developed during World War II by Doctor C. A. Lovell of Bell Laboratories. These early operational amplifiers were constructed using vacuum tubes and similar discrete components, and the intended purpose of these early op amps was for use in *analog* computers. Several analog computers using vacuum-tube op amps were actually built and put to work in aircraft and missile design. The term ''operational amplifer'' comes from the fact the circuit was designed to perform mathematical ''operations.''

All op amps—both vacuum tube and IC—are able to ''compute'' because of a characteristic mentioned in Chapter 1. All op amps amplify the *difference* between two signals applied to their inputs. Figure 2-1 shows the schematic symbol for an op amp (this symbol was introduced in Figure 1-5). The input signals applied to the inputs of an op amp can be dc or ac, and

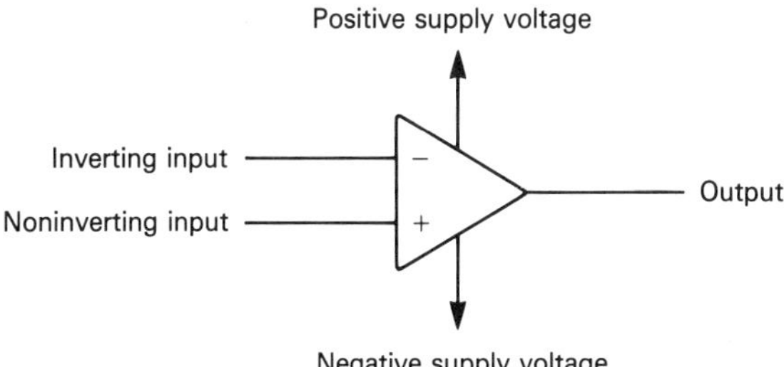

Figure 2-1 Schematic symbol for an operational amplifier.

the output signal waveform will depend on the waveform(s) of the input signal(s). Since the output is the difference between the input signals, an op amp can "subtract" one signal from another and return the difference. With a few external components, the same op amp can "add" together input signals and return their sum. Division can be performed as repeated subtractions, and multiplication is possible through successive additions. If the input signals are equal, the signals cancel and produce no output, or zero output. These concepts formed the basis of early analog computers.

Since the operational amplifier shown here amplifies the difference between the two voltages appearing at its inputs, it does not have a ground connection point on the IC itself or in its internal equivalent circuit. This is in contrast to almost all discrete transistor amplifiers and even digital ICs. A common ground connection point can be introduced to op amp circuits using external components if needed, and some op amp ICs are designed with a common ground point.

Although the inverting and noninverting inputs are designated by the − and + symbols, these have nothing to do with which signal polarities can or cannot be applied to them. Instead, they refer to what happens to a signal applied to a particular input. The noninverting input (+) simply means that the signal applied to it—whether of positive or negative polarity—retains its original polarity. The inverting input (−) means that the polarity of the input signal applied to it is *reversed.* Thus a positive-polarity signal (+) applied to the inverting input becomes a negative-polarity signal (−); by the same token, a negative-polarity signal applied to the inverting input becomes a positive-polarity signal. (Do not confuse the input "polarities" with the dual-polarity voltages necessary to power many op amps.)

Figure 2-2 shows how these inputs can be used to produce an output of the desired polarity. The *noninverting mode* produces an output which has the same polarity as the input; note that the inverting input is grounded. In the *inverting mode,* the output signal's polarity is the reverse of the polarity of the input signal. The inverting mode is very similar in concept to the NOT logic gate introduced in Chapter 1, and an inverting-mode op amp circuit can indeed function as a NOT gate if a digital signal is used as the input. Since the output of an op amp is the difference between its input signals, the output of both of these circuits is the difference between the input signal

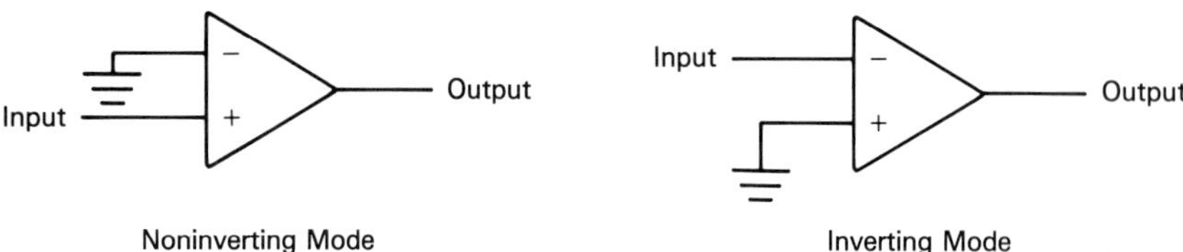

Figure 2-2 Inverting and noninverting modes of operational amplifier operation.

and ground (0 V). (Note that the dual-polarity power supply connections have been omitted here; that will be the general practice throughout the remainder of this chapter.)

If the input of an op amp is not being used, it should be grounded. An unused, ungrounded input is said to "float" and will often result in improper operation. Similary, an input that does not receive the input signal it is supposed to receive will also float. If you have trouble getting an op amp circuit to function properly, check to see if an unused input is grounded or if both inputs are receiving the proper input signals. (You will find this to hold true for most other ICs, digital and linear. An input must receive a signal or be grounded or "held" at a specified logic state.)

One interesting point about the inputs of an op amp is the remarkably low current they draw, often on the order of a few picoamperes. The amount for any op amp is so low that for all practical purposes the inputs of an op amp can often be considered as drawing no current at all.

IC op amps offer enormous *gain.* Gain is defined as the ratio between the output voltage and the input voltage. The gain offered by many op amps can be staggering; ratios of 10,000:1 or 100,000:1 are common. Usually, the input-signal part of the gain ratio is understood to be 1 and deleted, so you will see op amps described as having gains of 10,000 or 100,000. This means that exceedingly weak voltages can be amplified to the point where they can drive other circuits or headphones, speakers, meters, light-emitting diodes (LEDs), and so on. Theoretically, this means that a signal of 1 mV could drive an op amp to produce an output of over 50 V. However, the output voltage cannot exceed the power supply voltages applied to the op amp.

The gain of an op amp is independent of its supply voltage; increasing the supply voltages from 9 V to 12 V or higher will have no effect on its gain. (Of course, exceed its maximum supply voltage rating and you can decrease its gain to zero by destroying the op amp!)

The gain of an op amp can also be *linear.* This means that the amplified output of an op amp *can* be an accurate reproduction of the weaker input signal. Figure 2–3 shows a graphical representation of the relationship between input and output voltages when an op amp is operating in a linear fashion. The output voltage is proportional to the input voltage. If the relationship between input and output voltages is not linear, the result is a badly distorted (but amplifed) output that is not recognizable as the product of the input. Without linear amplification, it would be impossible to amplify most audio or radio signals.

In the preceding paragraph, it was stressed that an op amp *can* operate in a linear fashion. It does not do so "naturally," however. The op amps depicted in Figures 2–1 and 2–2 are operating in what is known as the *open-loop mode.* In this mode, the op amp is going "full blast" and its gain is close to its theoretical limit. Even the most minute signal can produce maxi-

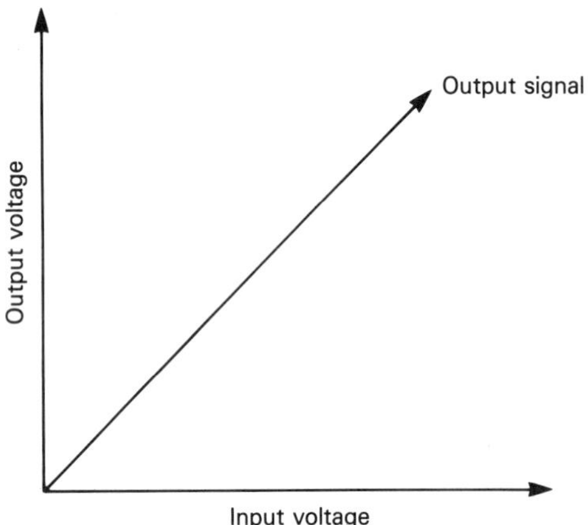

Figure 2–3 Graph depicting operation of a linear amplifier.

mum gain in the op amp; less than 0.001 V of input (the signal at one input with the other grounded or the difference between two input signals) can produce maximum output from the op amp. The result is that all input signals produce the same output level, and the output is a distorted mess bearing no apparent resemblance to the inputs that produced it.

As you might expect, op amps are rarely operated in the open-loop mode. There is one type of op amp circuit which does use this mode, however, and it is known as the *comparator.* A comparator operates as an on/off switch. For example, suppose that some method is needed to detect when two voltages differ from each other. If both voltages are used as the inputs to a op amp comparator, the comparator will produce an output if they vary by the slightest degree. A common comparator application is to supply a voltage, known as a *reference voltage,* to one input and a voltage to be monitored to the other. In this way, the output of the op amp can be turned off or on depending on whether the monitored voltage exceeds or falls below the reference voltage. Special op amps intended for use as comparators have been developed, and applications of one common type will be discussed later.

However, for the vast majority of op amp applications some sort of "brake" must be applied to the gain of the op amp so that it will operate in a linear fashion. The way to control the gain of an op amp is through *feedback.* Some of the output of the op amp is taken and "fed back" into one of the inputs through a *feedback resistor.* Figure 2–4 shows an *inverting amplifier* whose gain is controlled through a feedback resistor. Such an op amp is said to be operating in the *closed-loop mode.*

Note that the input signal is applied to the inverting input and that the output is the opposite polarity of the input, hence the name "inverting amplifier." The gain of this circuit is determined by the values of the feed-

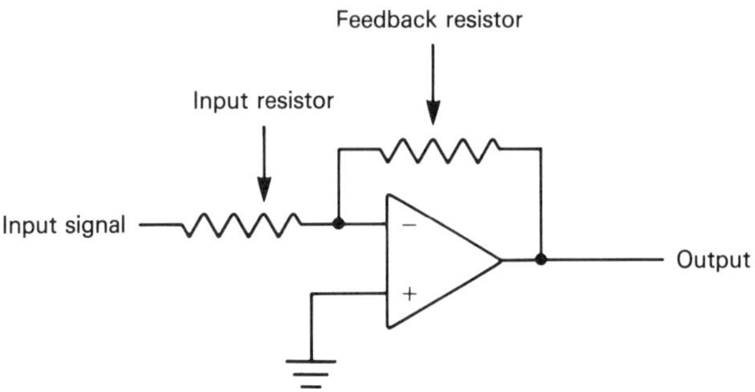

Figure 2–4 Inverting operational amplifier with feedback and input resistors.

back and input resistors. The gain is demonstrated by dividing the value of the feedback resistor by the value of the input resistor, and negating the result, or

$$\text{gain} = -\frac{\text{feedback resistor}}{\text{input resistor}}$$

For example, let's assume that the feedback resistor has a value of 100,000 ohms (Ω) and the input resistor has a value of 1000 Ω. The gain is found by 100,000/1000, giving a gain of 100. Negating the result produces a final answer of – 100. (The "– 100" figure does not imply a loss; instead, it means that the output is the opposite polarity of the input signal.)

Although fixed-value resistors have been used in this example for simplicity, there is no need for them to be so. Potentiometers can be (and often are) substituted to give variable gain over a wide range.

Figure 2–5 shows what happens to both sine-wave (analog) and pulse (digital) signals during inversion by the circuit in Figure 2–4. The waveforms are not altered in their amplitude, duration, or wavelength; instead, they are "turned upside down," so that the positive peak of the input signal happens when the output signal is at its negative peak.

The *noninverting amplifier* amplifies like the circuit in Figure 2–4, but its input is not reversed at the output. Figure 2–6 shows a noninverting amplifier; spend a few moments and compare it to the circuit in Figure 2–4. (Note that they look almost like "inverted" versions of each other.) As with the inverting amplifier, the gain is determined by the values of the feedback and input resistors. The formula for this is

$$\text{gain} = \frac{\text{feedback resistor} + \text{input resistor}}{\text{input resistor}}$$

For example, assume that the value of the feedback resistor in Figure 2–6 is 10,000 Ω and the input resistor is 1000 Ω. The gain of the circuit would be (10,000 + 1000)/1000, or 11,000/1000 = 11.

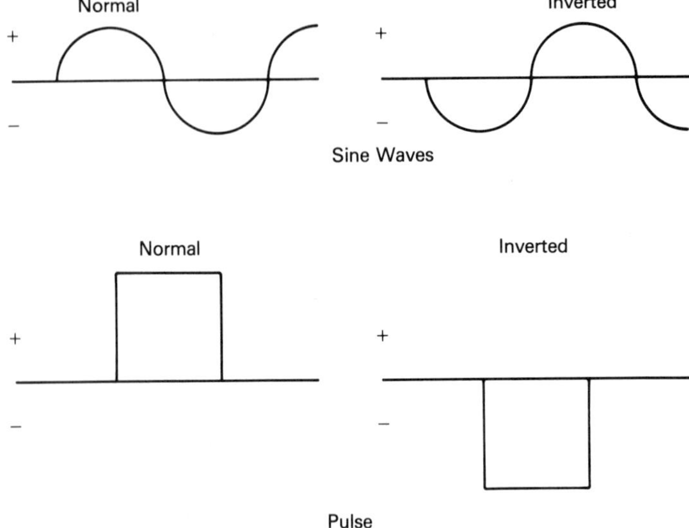

Figure 2–5 Inverted and noninverted forms of sine and pulse (square) waveforms.

These formulas for determining the gain of inverting and noninverting amplifiers work for almost all op amps. The ultimate limits to gain can be found by referring to the data sheets for the op amp in question.

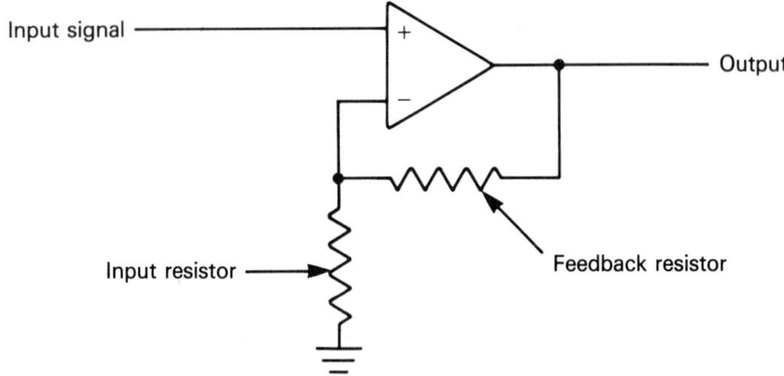

Figure 2–6 Noninverting operational amplifier with feedback and input resistors.

OP AMP SPECIFICATIONS

Gain is not the only important op amp specification. In fact, most op amps have more than enough gain for most applications, so other specifications may become more important in deciding which op amp is best for a particular purpose. Most of the time you will not have to concern yourself with them, but they do show up in magazine articles and op amp reference material (such as manufacturers' data sheets), so let's discuss the more common specifications.

Important specifications for op amps used with ac input signals are *bandwidth* and *unity-gain frequency.* These both indicate the upper limits of an op amp's ability to handle ac signals. The unity-gain frequency is that frequency at which the gain of the op amp drops to 1; the input signal voltage equals that of the output signal voltage at this frequency. An op amp is useful for amplifying ac signals only at frequencies *below* the unity-gain frequency. The bandwidth is the frequency range over which the op amp is useful, from the maximum (the unity-gain frequency) down to the lowest frequency, which is often 10 to 100 times lower than the unity-gain frequency. Bandwidth varies in the open-loop and closed-loop modes, but is normally measured and expressed in the closed-loop modes.

Slew rate is determined when an ac sine-wave input is used to "overdrive" an op amp so that the output is a triangular waveform rather than a sine wave. The slope of the triangular output is the slew rate. A more practical working definition is that slew rate is the rate of change in the output of an op amp, measured in volts per microsecond (V/μs), when the gain of the op amp is unity (or 1). A typical value is 1 V/μs, although some op amps have been designed for values over 100 V/μs. The slew rate is a measure of how "fast" an op amp is and the ability of its output to repond rapidly to changes in its input.

The inputs of an op amp are not precisely "balanced." Even if no input signal is being applied, a small output voltage may be produced. To prevent this, many op amps include connections for two *offset voltages* (one for each input), which can be used to set the output to zero when no input signal is present. The *input offset voltage* is used to describe those voltages used for zero output, with typical and maximum values given. Fortunately, for most applications no offset voltages need to be applied.

Another specification related to inputs is the *common-mode rejection ratio* (CMRR). Many applications require that the op amp not amplify (*reject*) a common signal applied simultaneously to both inputs. CMRR measures the op amp's ability to reject such common signals.

The *input resistance* is normally measured between the two inputs, and the upper limit can be several megohms in popular op amps. Some op amps that have integrated field-effect transistors can have input resistances which are dramatically higher than this figure.

Fortunately, for many applications you will not have to concern yourself with these more exotic specifications. The only crucial specification to observe is *supply voltage.* An op amp must always be operated within its specified supply voltage; if the maximum supply voltage is exceeded, the op amp may be destroyed.

The circuits in the rest of this chapter have already been designed using popular and readily available op amps. Most will also function if other op amp ICs are substituted; however, be sure to determine the pin connections, supply voltage, and other vital parameters for the device you intend to substitute. The manufacturer's data sheet is the best source for this information.

CIRCUITS USING THE 741 OP AMP

As mentioned earlier, the most popular op amp IC in use today is the 741, first introduced by Fairchild. This device is currently available from over a dozen different manufacturers and is inexpensive; two or three 741 ICs can usually be obtained for less than a dollar.

The 741 contains the integrated equivalent of 20 transistors and 11 resistors. Its nominal supply voltage requirements are − 15 V and + 15 V, with a maximum supply voltage of − 18 V and + 8 V, although it will work fine from the dual-polarity 9-V supplies shown in Figures 1–11 and 1–12. It is capable of voltage gains up to 200,000 and its unity-gain frequency is approximately 1500 kilohertz (kHz). Its slew rate is modest but is suitable for many applications. The only operating restriction that you should observe is to keep input voltages below supply voltages when the supply voltage is under 15 V.

Figure 2–7 illustrates the connections and internal diagram of the 741. As you can see, the inverting input is at pin 2 and the noninverting input is at pin 3, with output taken from pin 6. The negative-polarity supply voltage is applied to pin 4 and the positive-polarity supply voltage is applied to pin 7. The offset null points are at pins 1 and 5; for the circuits in this chapter using the 741 you can forget about them. (For applications where offset voltages are required, the usual procedure is to connect a potentiometer across pins 1 and 5, with the wiper connected to a voltage source.)

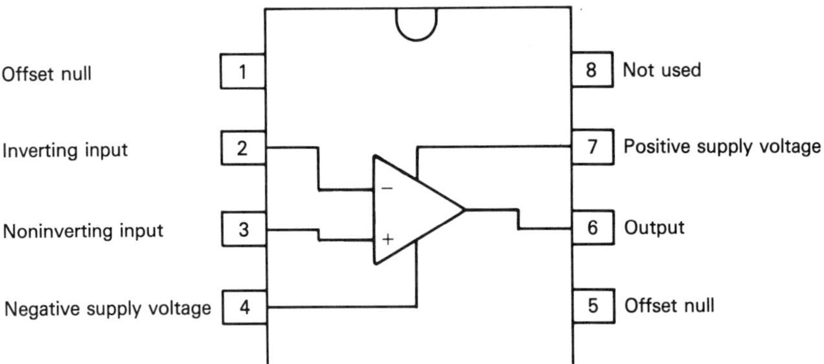

Figure 2–7 Pin connections and internal diagram for 741 operational amplifier.

The remaining pin, pin 8, is labeled "not used" in Figure 2–7. It is not connected to the circuit in the package in any way, and is present simply because the standard DIP housing the IC has eight pins. One apocryphal story holds that early IC designers wanted to label this pin "optional ground," to satisfy those engineers who were bothered by the lack of a ground connection point on the 741!

Figure 2–8 shows a model for a basic inverting amplifier whose gain you can alter to your specific needs. The gain of the circuit is determined by the formula

$$\text{gain} = -\frac{R2}{R1}$$

Unlike the idealized model of an inverting amplifier given in Figure 2–4, the noninverting input is grounded through a resistor, R3. R3 is often necessary because, as you might expect, the real world is more complex and variable than the model suggests. When R3 is used and grounded as shown in Figure 2–8, it helps ensure that both inputs have the same resistance and that zero input signal causes zero output signal. The value of R3 is determined by the formula used to determine the value of R1 and R2 in parallel, or

$$R3 = \frac{R1 \times R2}{R1 + R2}$$

R3 is not necessary in every instance; you may be able to connect the inverting amplifier directly to ground without adverse results. Should you have trouble getting it to function correctly, however, the addition of R3 may help.

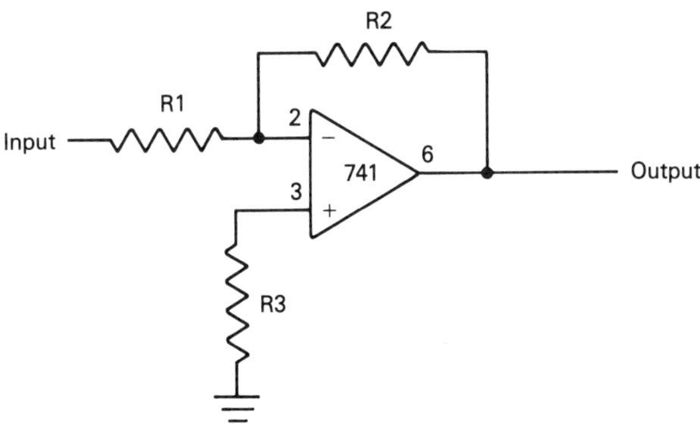

Figure 2–8 Basic inverting amplifier using the 741.

Figure 2–9 shows a basic noninverting amplifier built around the 741. The gain of this circuit is determined in the same way as the gain for the model noninverting amplifier shown in Figure 2–6, with the formula for Figure 2–9 being

$$\text{gain} = 1 + \frac{R2}{R1}$$

Thus, if the value of R2 is 36 kilohms (kΩ) and the value of R1 is 4 kΩ, the gain of the amplifier is 10.

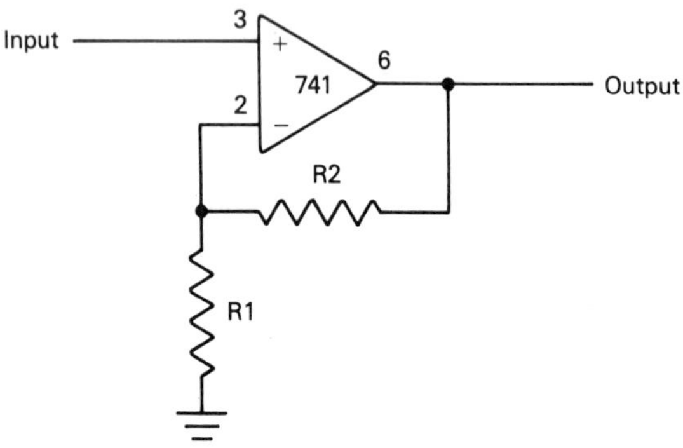

Figure 2–9 Basic noninverting amplifier using the 741.

The circuit in Figure 2–9 is best suited for dc input signals. If it is used with ac input signals, it may be a good idea to set the gain of the op amp to unity (1) at a dc input. This can be done by placing a capacitor between R1 and ground. The value required for the capacitor is usually large, often on the order of 4.7 microfarads (μF) or greater. The exact value is determined experimentally until best performance is achieved.

Both noninverting and amplifier circuits are subject to excessive feedback through the feedback resistor (R2 in Figures 2–8 and 2–9). When this happens, the op amp *oscillates* and the resulting output is a badly distorted waveform. This can be remedied by placing a capacitor of approximately 100 picofarads (pF) across the feedback resistor. If the oscillation persists, the value of the capacitor should be increased. This approach will decrease somewhat the high-frequency gain of the circuit but usually not to a significant degree.

There are situations in which unity gain is desired. Figure 2–10 shows

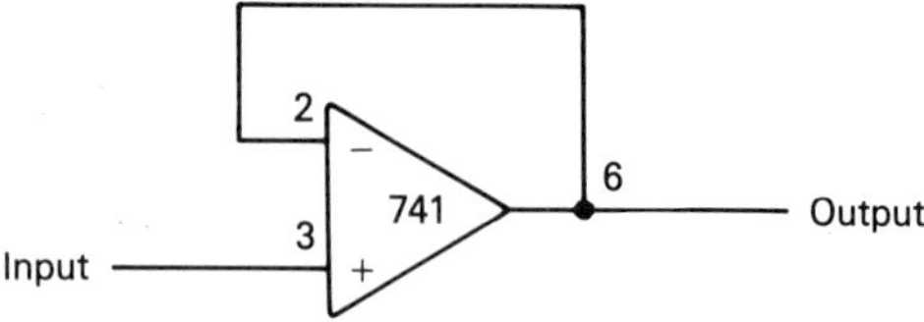

Figure 2–10 Unity-gain buffer amplifier built around the 741.

a unity-gain noninverting amplifier; the input voltage equals the output voltage. An amplifier of this type, known as a *buffer amplifier,* is used to "isolate" various circuits from each other.

In other cases, unity gain but an inverted output may be desired. Figure 2-11 shows such as *inverter.* This can serve as a buffer (provided that the inverted output is not a problem) but can also be used in the same manner as a NOT logic gate.

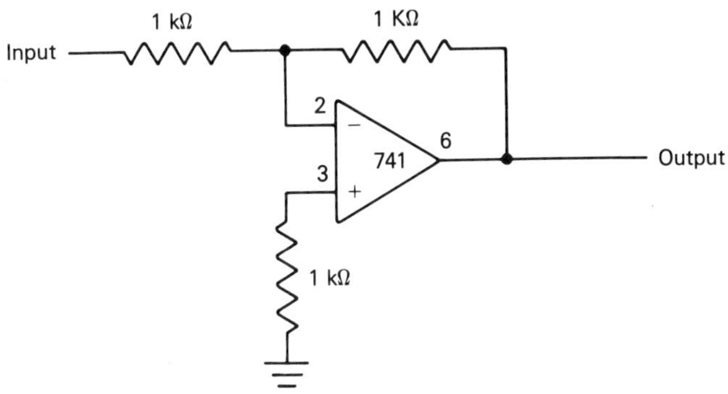

Figure 2-11 Unity-gain inverter amplifier using the 741.

Figure 2-12 shows an open-loop mode comparator. A steady reference voltage is applied to the inverting input and the input voltage to the noninverting input. When the input signal is less than the reference signal, the output of the op amp is zero. When the input signal exceeds the reference signal, the output of the op amp leaps to maximum. The resulting output signal has the same polarity as the input signal, so Figure 2-12 shows a *noninverting comparator.* It is also possible to create an *inverting comparator.* To do so, the reference signal is applied to the noninverting input and the input signal is applied to the inverting input.

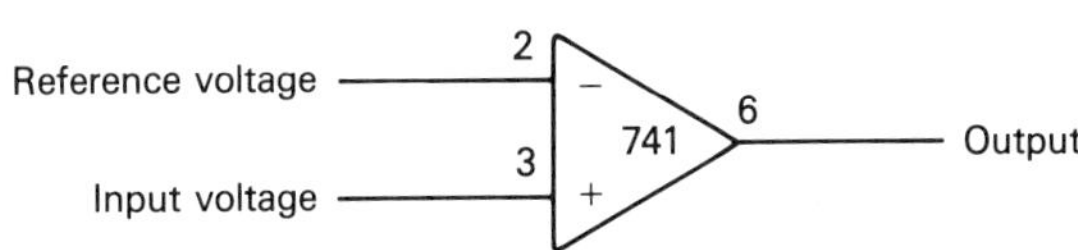

Figure 2-12 The 741 configured as a "bare bones" comparator.

As mentioned earlier in the chapter, the operational amplifier circuit was originally developed for analog computation. Examples of these "original" op amp applications circuits can be implemented using the 741. Figure 2-13 shows a *difference amplifier,* which performs subtraction of two signals. This circuit produces an output equal to the following formula:

$$\text{output voltage} = \text{input } 2 - \text{input } 1$$

The polarity of the output signal is the same as that of the input signal in this circuit.

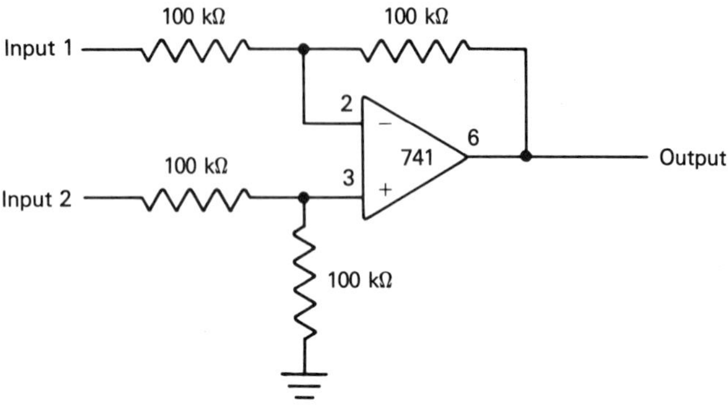

Figure 2–13 Difference amplifier.

Other op amp circuits add the two input voltages together and their sum appears as the output. This is known as a *summing amplifier,* and Figure 2–14 shows a schematic for such a circuit. Although only two inputs are shown, additional inputs may be added to the inverting input (pin 2) as long as each input is connected through a 10-kΩ resistor. The sum of all inputs should not exceed the supply voltage; good practice is to keep the sum of all inputs approximately 1 V under the supply voltage.

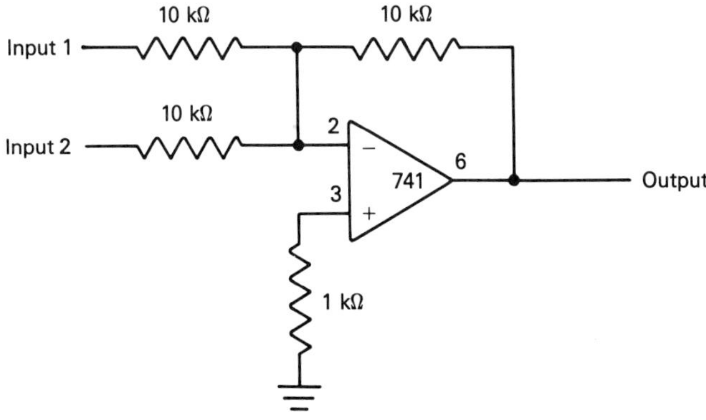

Figure 2–14 Summing amplifier; the output is the sum of the input signals.

MORE 741 APPLICATIONS CIRCUITS

The 741 may be used in oscillator circuits as well. Figure 2–15 shows the 741 used in a sine-wave oscillator whose output frequency depends on the values of R1 and R2. The values of R1 and R2 are identical; the greater the values of R1 and R2, the lower the oscillation frequency. Values used should be in the range 4.7 to 18Ω, with values of approximately 13Ω producing an output of around 1000 Hz. The exact frequencies you obtain for given values of R1 and R2 will depend on a number of factors, including the precision of the various resistors and capacitors used. The 100-kΩ potentiometer is adjusted until the circut "breaks into" oscillation. This circuit is a variation of the "twin-T" configuration.

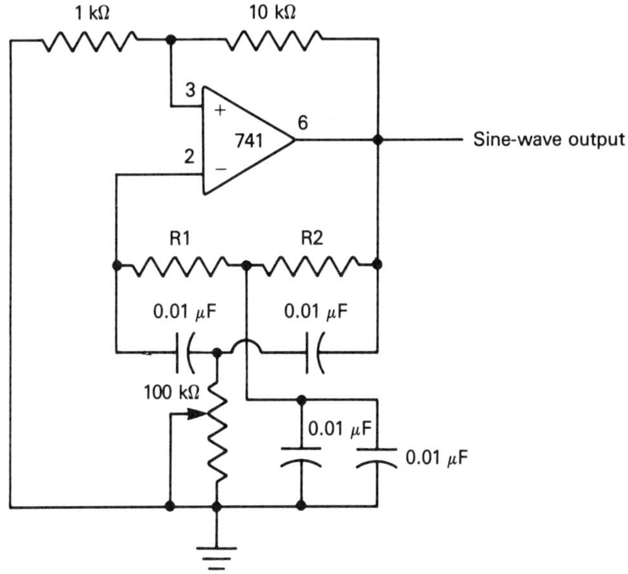

Figure 2–15 Sine-wave oscillator.

Figure 2–16 shows a square-wave oscillator built around the 741. This circuit is capable of operating over a wide range, from more than 10,000 Hz to less than 100 Hz. Pulse duration is controlled by adjustment of the 100-kΩ and 1-MΩ potentiometers. Note that unlike other circuits described so far in this chapter, Figure 2–16 uses a ground connection point and output is taken between the 741's output and the ground connection point.

This circuit can be modified in numerous ways, such as substitution of fixed-value resistors if a specific pulse rate is all that is needed. For example, the 10-kΩ potentiometer connected to pin 3, together with the two 10-kΩ resistors in series with it, could be replaced by a single-value resistor (4.7 to 6.8 kΩ) connected between pin 3 and the supply voltage and another resistor of the same value between pin 3 and the ground connection point. The 100-kΩ potentiometer and 1-kΩ resistor between pins 2 and 6 could be replaced by a single resistance of approximately 100 kΩ and the 1-kΩ resistor and 1-MΩ potentiometer between pins 3 and 6 could be replaced by a single

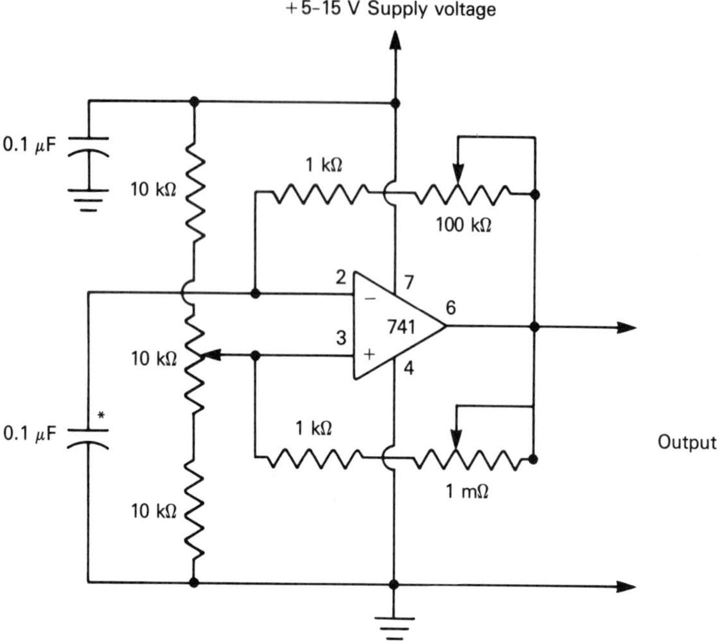

Figure 2–16 Square-wave oscillator.

resistance having a little more than one-fifth of the value of the new resistance between pins 2 and 6 (22 kΩ in this example). The output frequency in this case would be controlled by altering the value of the capacitor marked with an asterisk (*) in Figure 2–16. In this case, increasing the value of the capacitor will increase the duration of the pulses produced by the oscillator.

The circuit in Figure 2–16 can be more "touchy" than others in this chapter. Part leads should be as short as possible; in particular, the lead lengths between the two capacitors and the pins of the 741 must be kept short.

The term *integration* in electronics has a meaning borrowed from calculus, and refers to a circuit or process that produces an output proportional to the integral of the input signal. A term often used in conjunction with integration (and also borrowed from calculus) is *differentation,* which describes a circuit or process which produces an output voltage that is proportional to the rate of change in the input voltage. These definitions are easier to visualize than to describe in words, so Figure 2–17 shows the input and output waveforms for *integrator* and *differentiator* circuits. As you can see, an integrator is a circuit that takes a square-wave input and produces a triangular, "sawtooth" output, while a differentiator takes a triangular input and produces a square-wave output.

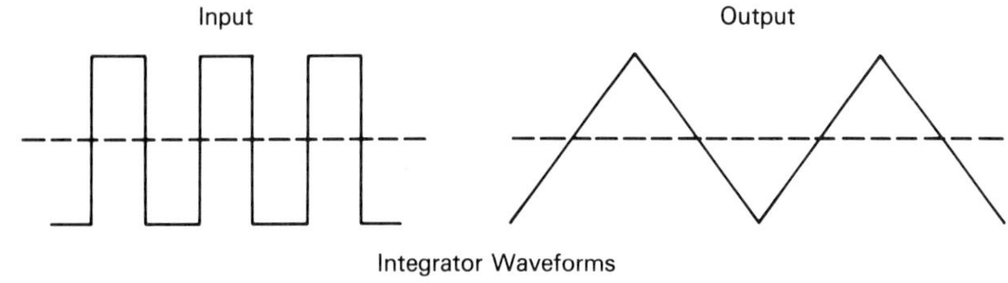

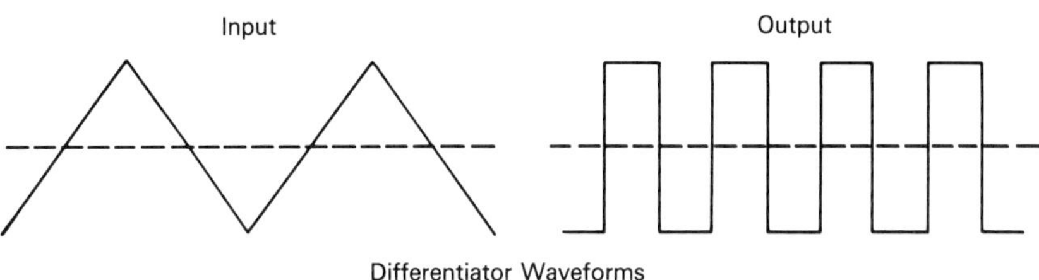

Figure 2–17 Input and output waveforms of integrator and differentiator circuits.

Figure 2–18 shows a "classic" integrator circuit built around the 741. The values of the components R1, R2, R3, and C depend on the frequency of the square-wave input. The value of R1 is approximately one-tenth of the value of R2; if R2 has a value of 560 kΩ, the value of R1 should be 5.6 kΩ. This can be expressed as

$$R1 = \frac{R2}{10}$$

The value of C is determined by the reciprocal of the value of R2 multiplied by the input frequency, or

$$C = \frac{1}{\text{input frequency} \times R2}$$

One can also determine the value of R2 if the values of C and the input frequency are known, as follows:

$$R2 = \frac{1}{\text{input frequency} \times C}$$

The value of R3 is determined by the formula for the value of two resistors in parallel:

$$R3 = \frac{R1 \times R2}{R1 + R2}$$

Similar design equations apply for the differentiator shown in Figure 2–19. The same relationship applies between R1 and R2, with R1 being

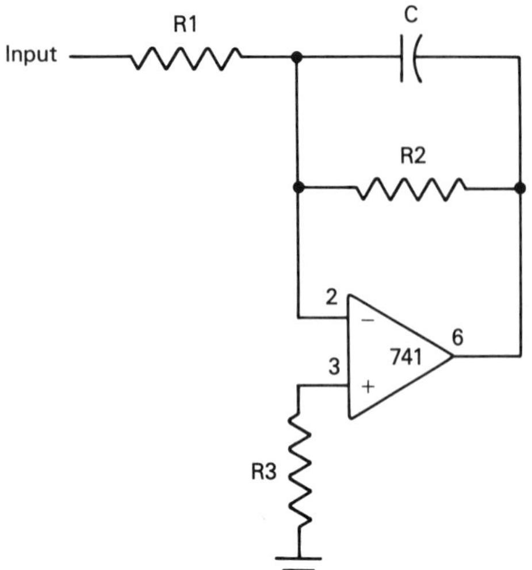

Figure 2–18 Integrator using the 741.

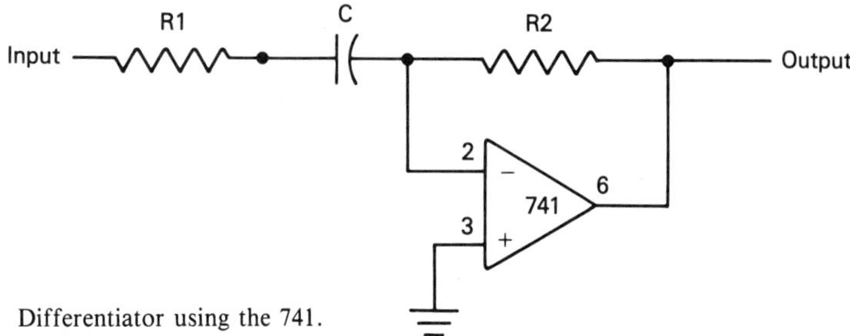

Figure 2–19 Differentiator using the 741.

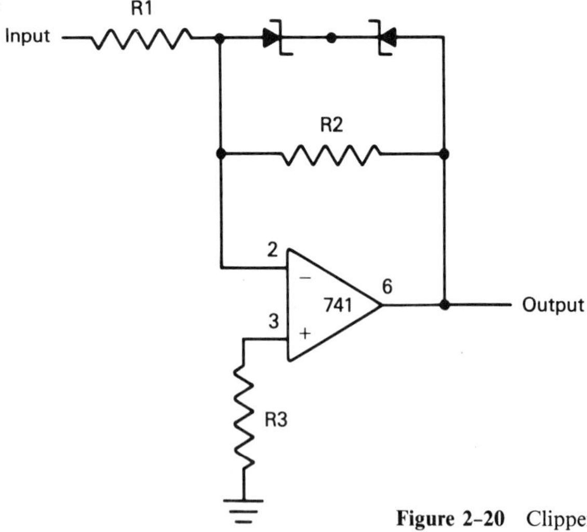

Figure 2–20 Clipper circuit with inverted output.

approximately one-tenth of the value of R2. The value of C can be found by the formula

$$C = \frac{1}{\text{input frequency} \times \text{R2}}$$

Similarly, the value of R2 can be found if the value of C is known, by the formula

$$\text{R2} = \frac{1}{\text{input frequency} \times \text{C}}$$

In addition to triangular waveform inputs, the differentiator is capable of accepting input in the form of square waves. When this happens, the output is a series of very "narrow" pulses.

Both the integrator and differentiator can be difficult circuits to get functioning properly. The values of the various resistors and capacitors can be crucial for proper functioning on a desired input frequency. A large number of resistors and capacitors of different values and patience can be a big help in "debugging" either circuit.

A cruder method of converting a sine-wave input into a more "squared" waveform is the *clipper*. As the name implies, the clipper "shaves" the round peaks off a sine-wave input and produces an output whose waveform exhibits flat peaks and gently sloping sides. Clippers are variations of conventional inverting and noninverting amplifiers, and follow similar rules for determining the gain for each. The gain of the *inverting clipper* shown in Figure 2–20 is found by the formula

$$\text{gain} = -\frac{\text{R2}}{\text{R1}}$$

with the value of R3 being equal to R1. In the *noninverting clipper* shown in Figure 2–21, the gain can be determined from the formula

$$\text{gain} = 1 + \frac{\text{R2}}{\text{R1}}$$

All diodes in Figures 2–20 and 2–21 are zener types with a 5-V rating.

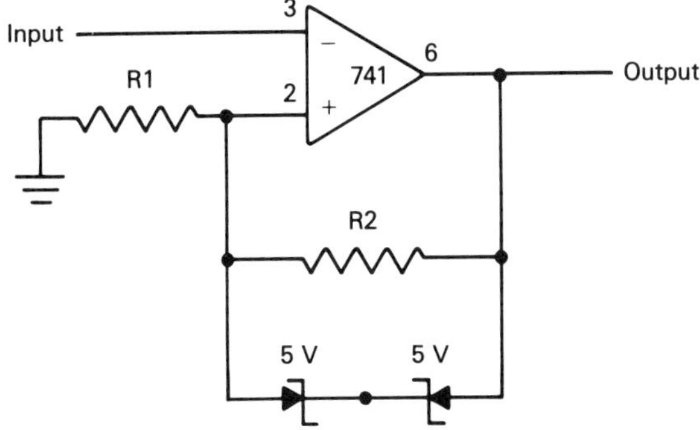

Figure 2–21 Noninverting clipper using the 741.

ACTIVE FILTERS

An *active filter* is a circuit that passes or rejects specified frequency ranges. Op amps can be used as active filters by adding components that tailor the frequency response of the op amp to the desired frequencies. Active filters are one of the most common uses of op amps.

Figure 2–22 shows the schematic for a *high-pass filter.* As the name suggests, the filter allows all frequencies *above* a certain frequency (the *cutoff frequency*) to pass without attenuation, while sharply rejecting all frequencies below the cutoff frequency. In the circuit illustrated, both R1 and R2 have the same value as those of C1 and C2. The cutoff frequency can be found by the following formula:

$$\text{cutoff frequency} = \frac{1}{(2\pi RC)}$$

where $\pi = 3.14159$, R is equal in value to R1 or R2, and C is equal in value to C1 or C2. The gain of the filter is determined by R4 divided by R3. The output at the cutoff frequency is 0.707 that of the circuit's maximum.

Some minor changes in the circuit produces a *low-pass filter.* A low-pass filter operates in the opposite manner from a high-pass filter, by allowing all frequencies *below* the cutoff frequency to pass without attenuation while rejecting all those above the cutoff frequency. Figure 2–23 shows a diagram for a low-pass filter. Note the similarity between it and the high-pass filter shown in Figure 2–22; the only changes are the transpositions of R1 and C1 as well as R2 and C2. As with high-pass filters, R1 is the same value as R2 and C1 is the same value as C2. The same formulas for determining the cutoff frequency and gain of the circuit given above for high-pass filters also apply to low-pass filters.

Another active filter design is that of the *bandpass filter.* This type of filter passes the most input signal at a specific frequency known as the *center frequency,* with all frequencies located more than about 100 Hz above or below the center frequency sharply rejected. The graph of the "response" of a bandpass filter is sharply pointed or "peaked." Bandpass filters are usually used to detect a specific tone or audio frequency in a multiple sine-wave signal.

Figure 2–24 shows a circuit for a bandpass filter. The center frequency is selected by adjusting the 1-kΩ potentiometer. The greater the resistance selected, the lower the center frequency. For example, a potentiometer setting of approximately 930 Ω gives a 1000-Hz center frequency, while a setting of around 130 Ω results in a center frequency of 2000 Hz. A fixed resistor can be substituted if only one bandpass frequency is needed.

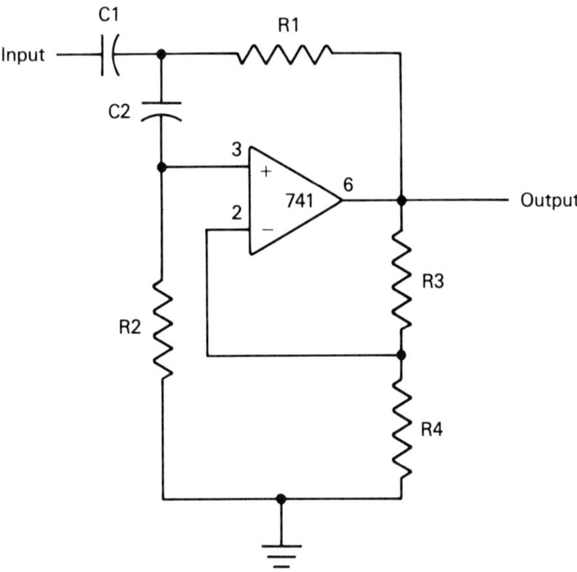

Figure 2–22 Basic high-pass active filter.

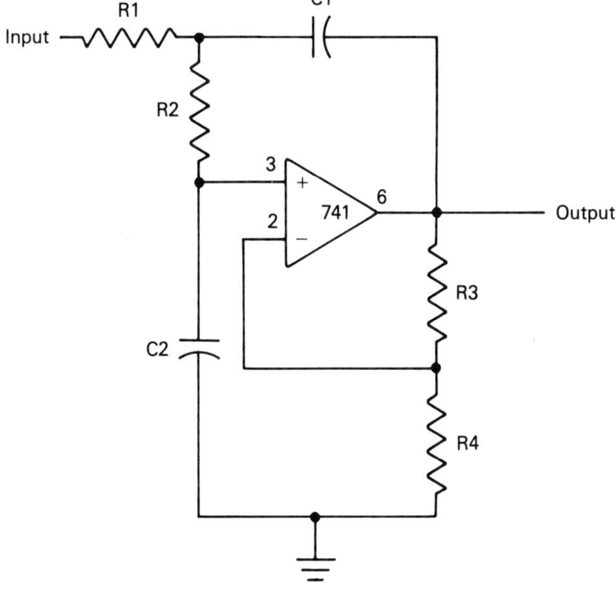

Figure 2–23 Basic low-pass active filter.

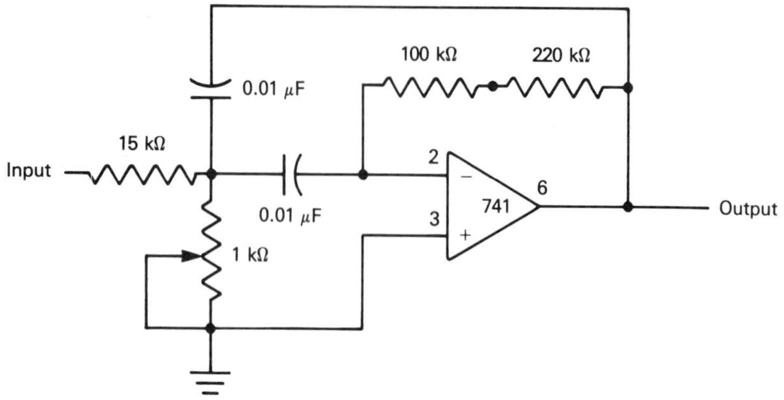

Figure 2-24 Tunable bandpass filter using the 741.

A useful variation of the bandpass filter for radio receivers is the *Q-multiplier,* shown in Figure 2-25. A Q-multiplier increases the ability of a radio receiver to receive a desired signal by having peak response on the fre-

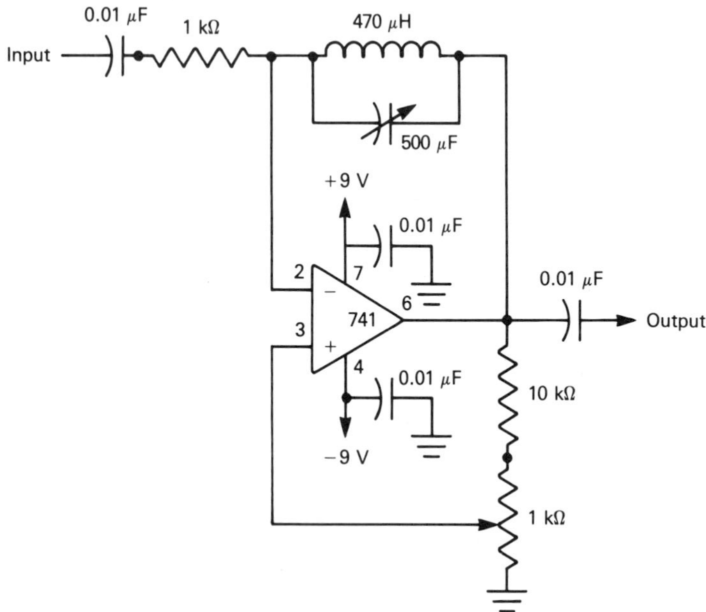

Figure 2-25 Q-multiplier circuit for radio communications.

quency of the desired signal while rejecting all other frequencies. In addition, a Q-multiplier can be used to reject a specific frequency on which an interfering signal is present. The circuit in Figure 2–25 was designed to be added after the single-sideband (SSB) filter in the 455-kHz intermediate-frequency amplifier stage of a receiver or transceiver. This circuit has a particularly "sharp" peak and is effective for receiving Morse code signals which are normally one tone (voice and music are composed of more complex waveforms). The "trimmer" capacitor across the 470-microhenry (μH) coil is used during circuit adjustment and testing; once the setting for proper operation is found, it is left alone. The center frequency of the Q-multiplier itself is determined by adjustment of the 1-kΩ potentiometer.

MISCELLANEOUS 741 APPLICATIONS CIRCUITS

The notion of a *current-to-voltage converter* might seem illogical, a bit like a "height-to-weight" converter for human beings! That is why it is also known as a *transresistance amplifier.* In operation, it produces an output voltage that is proportional to an input current. Figure 2–26 shows the schematic diagram for this circuit.

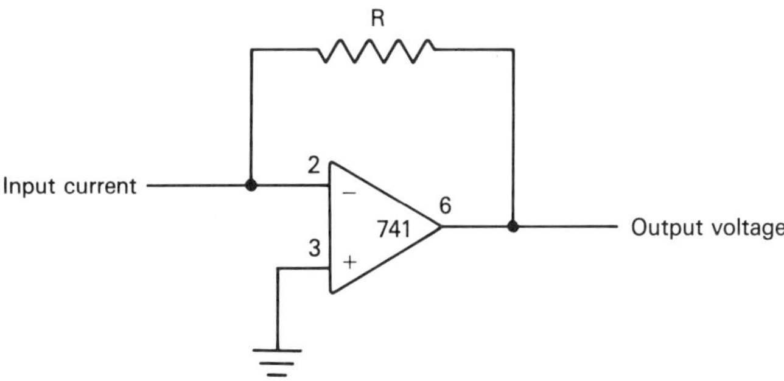

Figure 2–26 Basic current-to-voltage converter.

The gain of this circuit is controlled by the feedback resistor, labeled R. The output voltage is determined by multiplying the value of the input current by the value of R and then negating the result, or

$$\text{output voltage} = -(\text{input current} \times R)$$

The gain of the circuit is the output voltage divided by the input current;

it can also be found simply by negating the value of R. Thus, if R is equal to 1 kΩ, the gain of the circuit would be −1000.

Figure 2–27 shows the diagram for a *voltage-to-current converter,* also known as a *transconductance amplifier.* This circuit produces a current across the load resistance, RL, in proportion to the input voltage. The output current is "sensed" by resistor RS. The resulting voltage is then fed back in series with the input voltage.

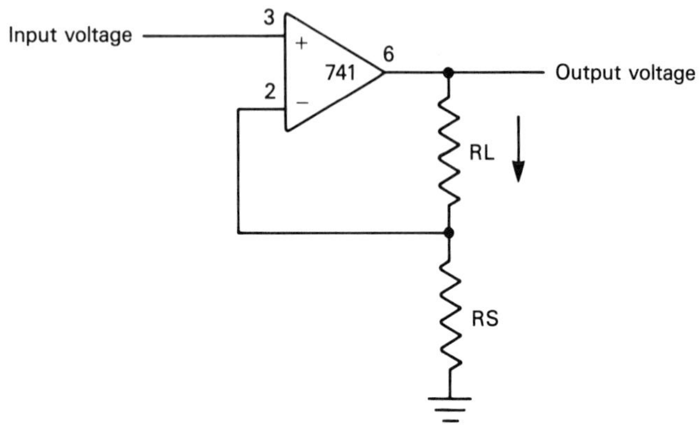

Figure 2–27 Voltage-to-current converter.

The output current is determined by dividing the input voltage by RS or by dividing the output voltage by the sum of RL and RS. The formulas for these are

$$\text{output current} = \frac{\text{input voltage}}{\text{RS}}$$

$$\text{output current} = \frac{\text{output voltage}}{\text{RL} + \text{RS}}$$

RL is normally not a resistor but another device, such as an LED or meter, which has a resistance value in a circuit. The value of RS is usually very small.

A variation of the comparator circuit discussed earlier is the *peak detector.* This circuit uses a capacitor to store the peak of the op amp's output. Figure 2–28 shows a peak detector circuit that incorporates a pushbutton switch, SW, so the 0.1-μF capacitor can be reset to zero after the stored voltage is read. A positive input voltage will charge the capacitor until the voltage across the capacitor equals the input voltage. When this happens, the capacitor stops charging until the input voltage increases beyond that stored by the capacitor; in this case, the charging of the capacitor resumes. The charged capacitor is "cleared" by pressing the pushbutton switch. The 1N914 diode prevents the capacitor from discharging through the op amp.

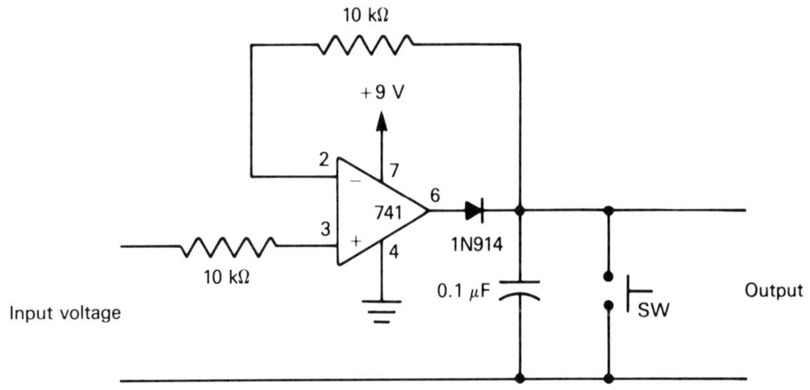

Figure 2–28 Resettable peak detector.

THE LM380 AUDIO POWER IC

Although extremely versatile, a general-purpose op amp such as the 741 is not the best choice for some situations. In particular, the amplification of signals at audio frequencies (10 to over 20,000 Hz) is an application for which other op amp ICs "optimized" for that purpose have been developed. The grandfather of all audio amplification ICs is the LM380, developed by National Semiconductor. Figure 2–29 shows the pin connections and internal diagram for the LM380.

Like conventional op amps, the LM380 has inverting and noninverting inputs. Unlike the 741, however, the LM380 has several functioning ground connection points and the input is referenced to ground. It requires only a single-polarity supply voltage, and the maximum supply voltage is 22 V, although it can operate satisfactorily at only 9 V. The LM380 can supply 3 to 4 W of power into an 8-Ω load such as a speaker.

An example of an LM380 application is the 2-W audio amplifier in Figure 2–30. This circuit is capable of operating from a supply voltage of +8 to +18 V and drives an 8-Ω speaker. This configuration draws less than 7 mA of current when no input signal is present.

Input signals for this circuit are applied through a standard phono jack input as shown in the schematic. The ground connection marked with an asterisk (*) is where pins 3, 4, 5, 7, 10, 11, and 12 should all be connected. Although the LM380 is capable of more than 2 W of output, the circuit in Figure 2–30 has very low distortion and operates over a very wide frequency range—in fact, over every frequency to which the human ear can respond. As such, this is an ideal circuit for music as well as for voice.

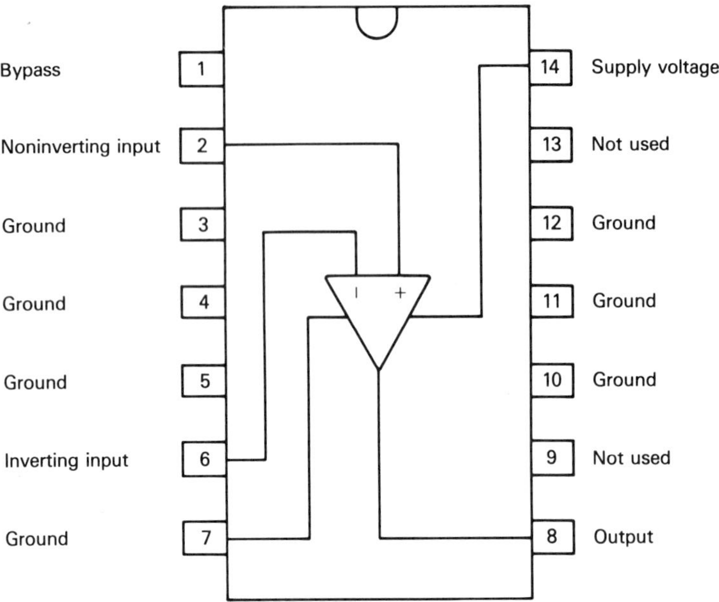

Figure 2–29 Pin connections and internal diagram of the LM380.

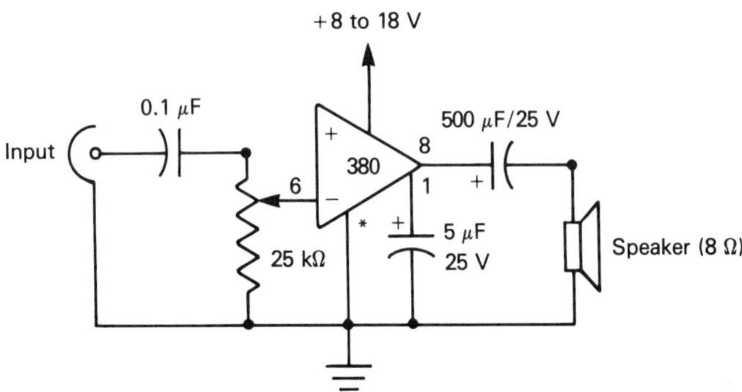

Figure 2–30 Two-watt audio amplifier using the LM380.

The LM380 was designed initially for applications in consumer electronics, such as portable record and tape players. Figure 2–31 shows a circuit designed for phonographs and record players conforming to the RIAA (Recording Industry Association of America) equalization standard for record playback. The RIAA standard calls for low frequencies to be attenuated (to prevent large undulations from breaking through phonograph groove walls) and for high frequencies to be boosted to improve the signal-to-noise ratio.

To achieve the RIAA equalization standard, this circuit has *inverse* frequency response and can be used in other situations where a similar frequency

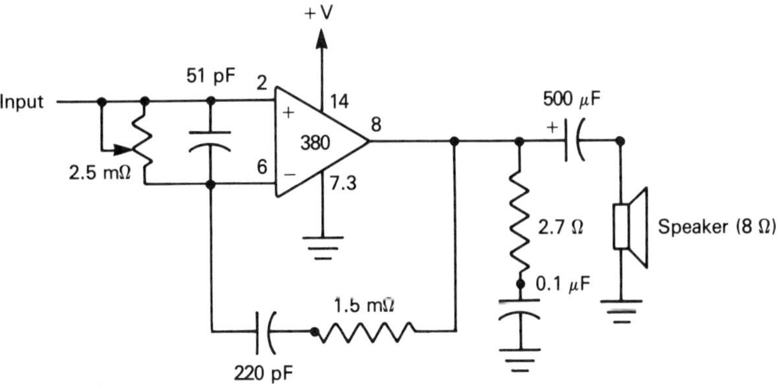

Figure 2-31 RIAA standard phono amplifier.

response is desired. The input is designed to be taken from a crystal cartridge. The 2.7-Ω resistor and 0.1-μF capacitor between the amplifier's output and ground help stabilize the circuit under high current loads.

The LM380 can be used in audio applications other than straight amplification. Figure 2-32 shows the circuit for a "no-loss" volume control. The LM380 eliminates signal attenuation found in conventional voltage-divider volume controls; the tone is controlled by varying the 2.5-MΩ potentiometer. The input is designed to be supplied by a source referenced to ground and the recommended supply voltage is + 18 V. Pins 3 and 7 are connected to ground.

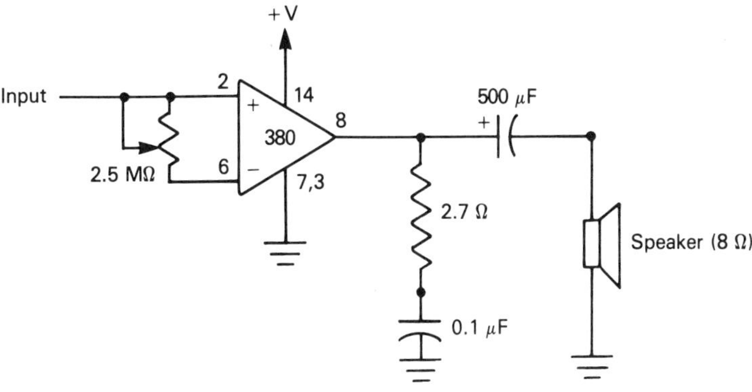

Figure 2-32 "Active" volume control.

Figure 2-33 illustrates the circuit for an intercom built around the LM380. The "L" and "T" next to the poles of the two SPDT switches stand for "talk" and "listen" and represent the positions they should be in for the desired function. The two speakers also serve as microphones, and both have 8-Ω impedances. The transformer can be any general audio type having 25:1 turns ratio.

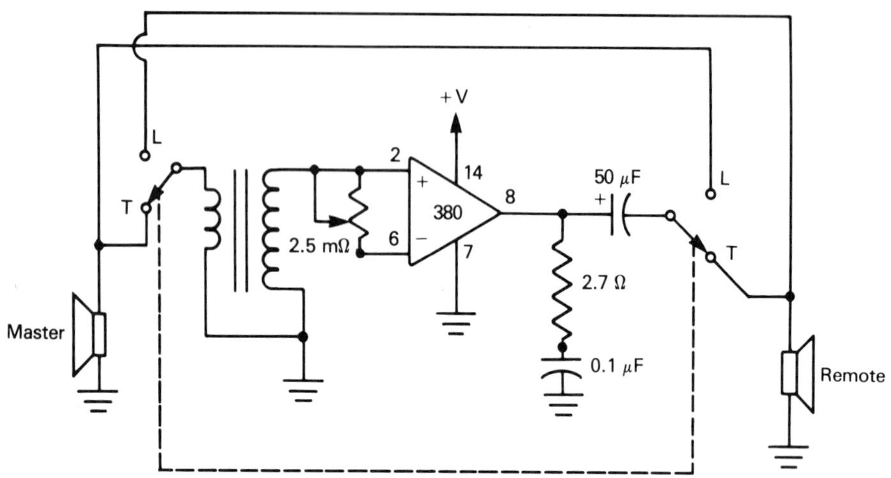

Figure 2–33 Intercom system using the LM380.

THE MC1458 DUAL-OP-AMP IC

Multiple op amps can be integrated onto a single slice of silicon and contained in one package. One of the most widely used "dual" op amps is the MC1458, developed by Motorola. Figure 2–34 shows its pin connections and internal diagram. While the two op amps share power supply leads, they operate completely independently.

Although the two op amps of the MC1458 are independent, they are normally used together in circuits. One application can be to produce summing and difference amplifiers whose outputs have the opposite polarity from their equivalents built around a single 741 (Figures 2–13 and 2–14). Figure 2–35 shows a summing amplifier using both "halves" of the MC1458. The output is again the sum of the two input voltages, but the polarity of the output is the same as the polarity of the inputs. The first op amp is used as a summing amplifier, while the second operates as an inverting buffer.

The difference amplifier in Figure 2–36 also produces an output that has the opposite polarity of its single op amp equivalent, thanks to the inverting buffer which follows the difference amplifier itself. The output of this circuit is found by subtracting the first input from the second input and then negating the result, or

$$\text{output} = -(\text{input 2} - \text{input 1})$$

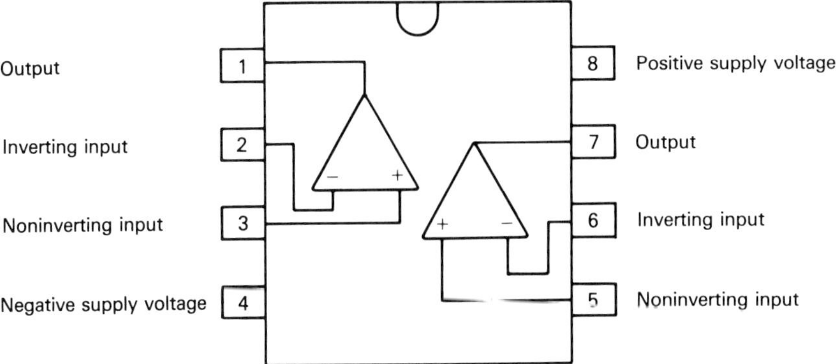

Figure 2–34 Pin connections and internal diagrams for MC1458 dual operational amplifier.

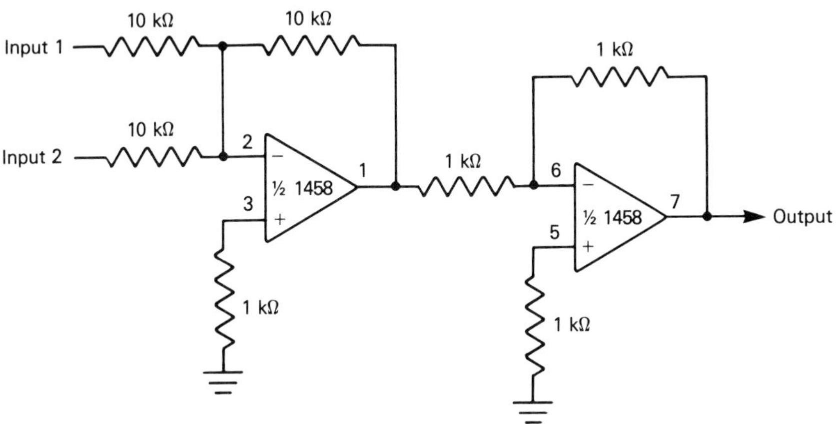

Figure 2–35 Summing amplifier using the 1458.

One useful application for dual-op-amp ICs is in *window comparators.* Figure 2–37 shows the schematic for one. The comparators mentioned so far have a single reference voltage, and their output depends on whether the input voltage is higher or lower than the reference voltage. By contrast, the window comparator has a high reference voltage and a low reference voltage. If the input voltage should fall into the range between the reference voltages, the comparator's output will change. If the input voltage is higher than the upper limit of the high reference voltage, or is lower than the lowest value of the low reference voltage, the output of the comparator is unchanged.

The circuit in Figure 2–37 normally has a high ("on") output. If the input voltage should fall into the range between the low and high reference

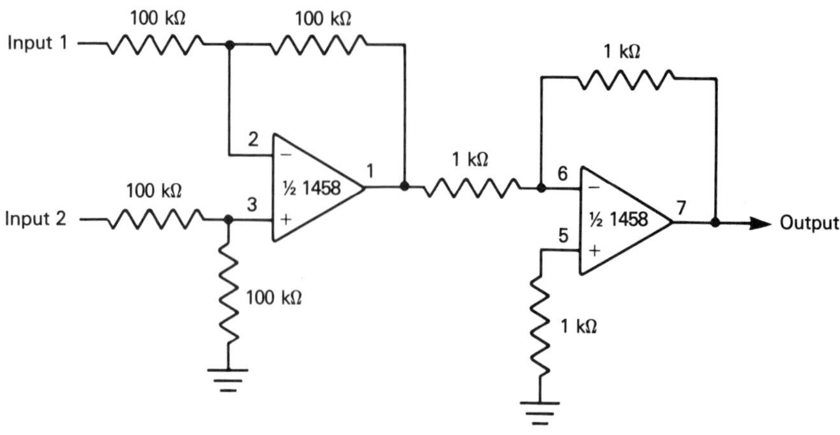

Figure 2–36 Difference amplifier using the 1458.

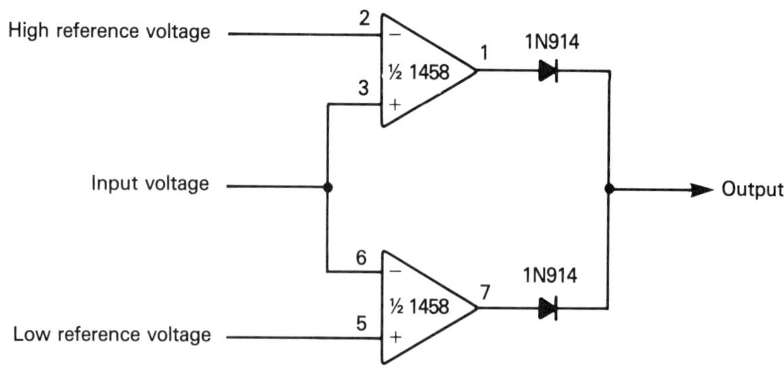

Figure 2–37 "Window" comparator using the 1458.

voltages, the output of the comparator will go low. The output of this circuit can be fed to an inverter if a reversed output polarity is needed.

A *peak voltage detector* monitors an input voltage and "stores" the maximum level it reaches in a capacitor. Figure 2–38 shows a dual-op-amp peak voltage detector that uses a 100-μF capacitor to "record" the maximum voltage input. The second op amp is used as a buffer between other circuits and the peak detector itself. The output of this circuit has the same polarity as the input.

Having two op amps conveniently available in one package permits building improved versions of circuits that previously were built using a single op amp. Figure 2–39 shows a pulse generator using both halves of the MC1458. This circuit has a much wider frequency range than the circuit in Figure 2–16. The frequency of operation is controlled by the value of the electrolytic capacitor labeled C; the greater the value of the capacitor, the

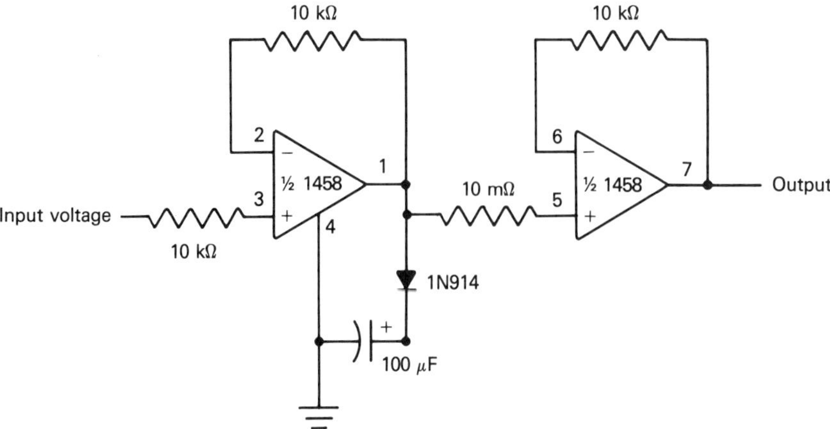

Figure 2–38 Peak voltage detector.

lower the frequency. For example, a value of 1 μF will produce an operating frequency of 8 Hz, while a value of 0.001 μF gives a frequency of almost 5900 Hz.

Other common capacitor values and their resulting output frequencies are 0.010 μF for 660 Hz and 0.1 μF for approximately 50 Hz. The amplitude of the output pulses are 5 V.

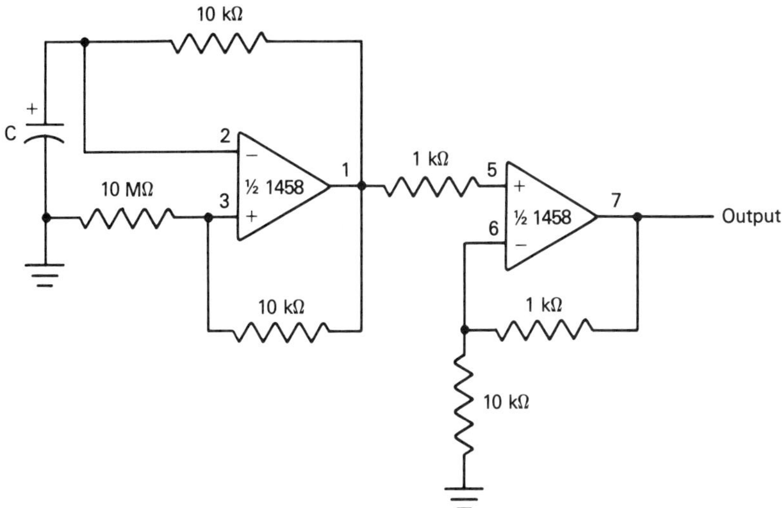

Figure 3–39 Pulse generator.

Figure 2–40 shows a *function generator* built around two MC1458 ICs. A function generator is a device capable of generating more than one wave-

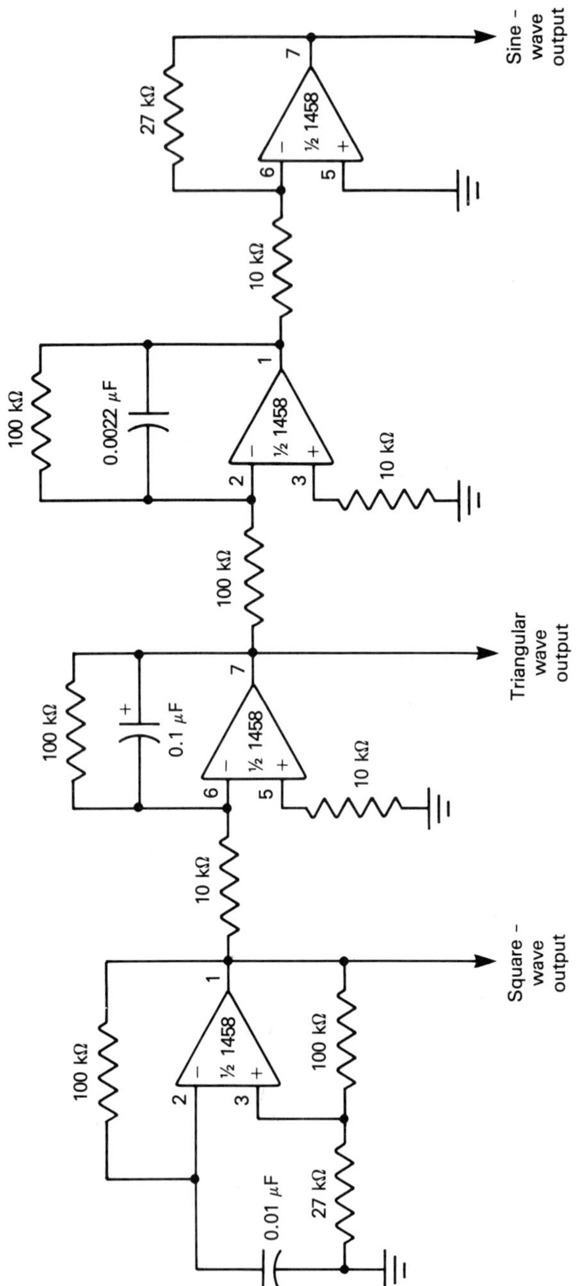

Figure 2–40 Tri-output function generator.

form, and the circuit in Figure 2–40 can produce square, triangular, and sine waves. The square-wave output has a peak (positive and negative voltages) amplitude of 7.5 V, while the peak amplitude of the triangular and sine waves is approximately 2 V. The operating frequency is approximately 1000 Hz.

In addition to dual-op-amp devices such as the MC1458, "quad"-op-amp devices, such as the LM324, are available. These contain four separate and independent op amps within the same package.

THE LM339 QUAD-COMPARATOR IC

The LM339 contains four op amps which have been optimized for use as comparators. Figure 2–41 shows the pin connections and internal diagrams for this device, which was first introduced by National Semiconductor.

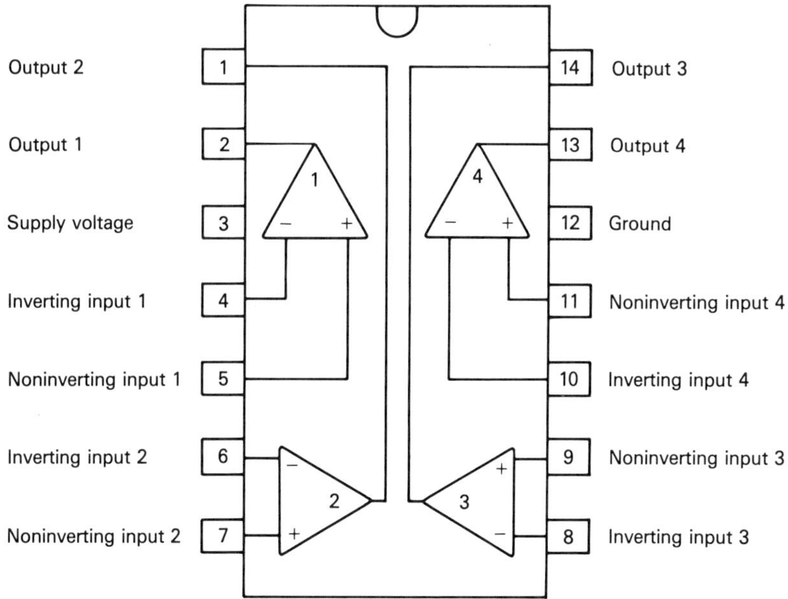

Figure 2–41 Pin connections and internal diagram of the LM339 quad comparator.

The four comparators in the package operate independently. The LM339 was designed to operate from a single power supply from 2 to 32 V dc; it is also capable of operating at half these levels from dual-polarity sources. It also includes a functional ground connection point at pin 12. Figure 2–42 shows one section of the LM339 configured to function as a comparator.

Perhaps the most obvious use of a comparator would be in digital systems, where the "on" or "off" state of the comparator would be used as an input for a digital IC of some sorts. The LM339 is particularly favored

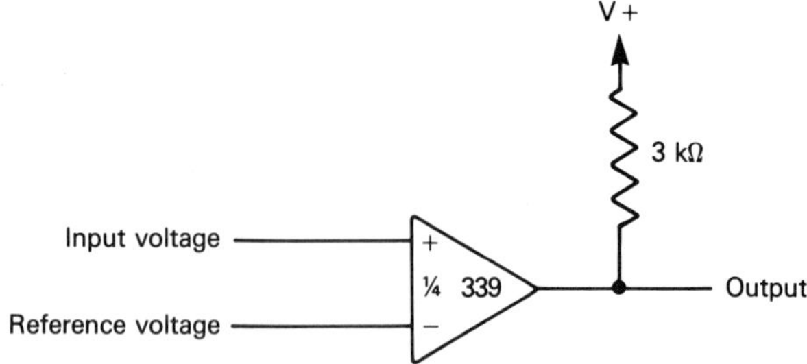

Figure 2-42 Basic comparator.

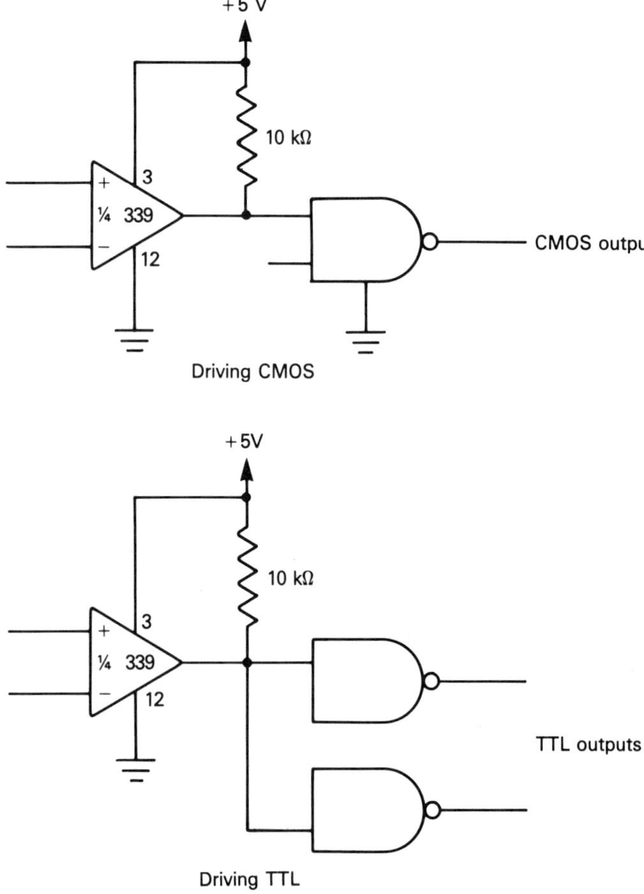

Figure 2-43 Driving TTL and CMOS logic devices from the LM339.

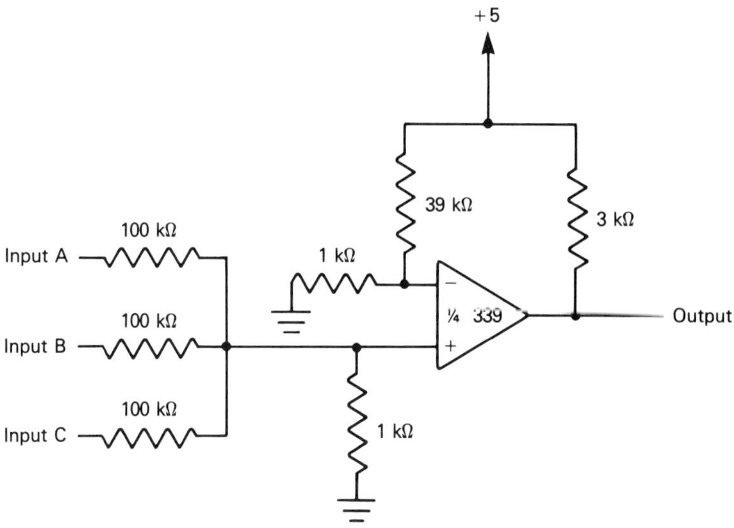

Figure 2–44 Three-input AND gate using the LM339.

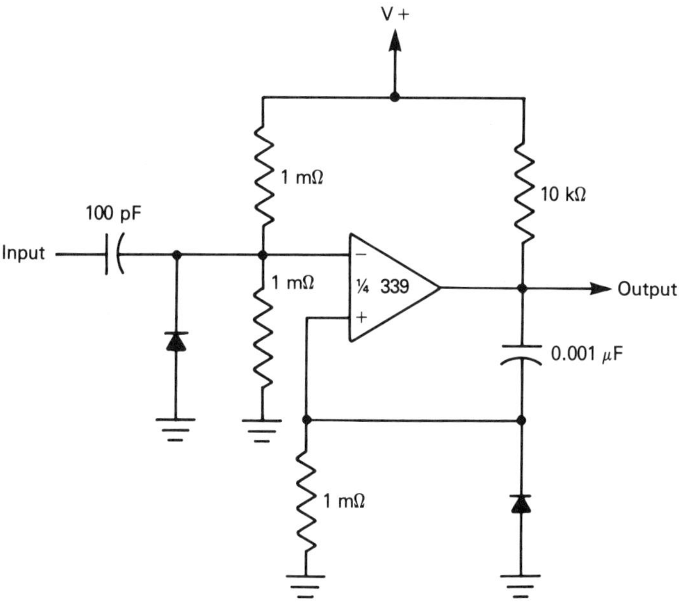

Figure 2–45 Monostable multivibrator.

for such applications because it can easily drive most major logic families without the need for extensive interfacing circuitry. Figure 2–43 shows how one section of the LM339 can be configured to drive CMOS or TTL family devices. The LM339 can also drive ECL and DTL family devices. When used to drive digital ICs, the supply voltage *must* be held to +5 V.

The characteristics of the LM339 that make it an excellent comparator and suitable for driving logic ICs directly also make it ideal for implementing logic circuits which have analog signal inputs. Figure 2–44 shows an AND gate built around one section of the LM339. The power supply must be kept at a constant +5 V and the output will be a constant +5 V. Like other AND gates, the output will be high when all the inputs are also high.

A *monostable multivibrator* is also known as a *one-shot multivibrator.* This is a circuit that will emit one pulse when "triggered" by an input signal. In the circuit in Figure 2–45, the circuit will emit a positive-voltage square-wave pulse lasting approximately 1 millisecond (ms) when the input voltage drops from a positive value to zero. Only one pulse is emitted, and the circuit is "reset" when the input voltage again rises from zero to a positive value. All diodes can be 1N914 or similar general-purpose devices.

ADDITIONAL TYPES OF OP AMPS

The op amps discussed so far in this chapter all use bipolar transistors. As mentioned in Chapter 1, additional types of transistors can be integrated onto silicon, and op amps using these different types of transistors are available. Although the basic application circuit configurations remain largely unchanged, op amps using transistors other than bipolar are often better suited for a particular application. Several op amps have been designed using *field-effect transistors* (FETs) in place of bipolar transistors. FETs have much higher input impedances than bipolar transistors, and a FET op amp will have very high input impedance as well. This means that FET op amps draw very little current from their input signal sources, and consequently have less effect on the input signal sources than does a conventional op amp. Unfortunately, early FET op amps were plagued by "drift" and instability due to temperature variations. This problem has largely been solved by op amps that use a combination of bipolar and FET transistors, known as *BiFET* op amps. In such devices, the FET transistors are used for input and bipolar transistors for the rest of the circuit.

Op amps have been designed using MOS transistors as well, with devices using combinations of bipolar and MOS transistors known as *BiMOS* op amps. A recent trend has been toward the development of op amp ICs fabricated using CMOS technology; several leading device designers in the IC industry have gone as far as to predict that CMOS op amps will eventually replace conventional bipolar types in the near future. The various MOS op amps offer low current consumption.

Other op amps—using bipolar and/or other device technologies—have been developed. These usually have some performance parameter enhanced or an additional feature so that they can be used in situations where more conventional op amps do not give the desired results. One specialized type of op amp is known as the *programmable op amp*. In this device, the internal operating currents are controlled by an external current applied to a "bias programming" pin. This allows such devices to operate over a wide range of supply currents and makes them well suited to operation from battery-powered supplies.

Various op amps are sometimes described as being *precision* or *high speed*. The term "precision" refers to how well the op amp can stay within its specified operating parameters. For example, in a comparator it refers to how well (or precisely) the op amp can detect when the input voltage differs from the reference voltage. It can also describe how well an amplifying circuit amplifies without distortion and noise or whether a summing or difference amplifier produces a correct output voltage. Using a precision op amp in a circuit does not mean that the circuit itself instantly becomes more "accurate," however; other components, such as resistors, must be of greater precision (such as 1%) in terms of their actual values. Using a precision op amp in a circuit with 20%-tolerance resistors is a wasted effort! A "high-speed" op amp is one that has a very high slewing rate without adverse side effects such as distortion or instability. As with precision op amps, other components, such as resistors and capacitors, need to be "optimized" to get the most out of the device's speed capabilities.

Precision versions of some popular conventional op amps are available, as are some devices that combine the attributes of high speed and precision. As you probably expect, precision and high-speed op amps are more expensive than their "normal" counterparts. However, these devices have wide use in such applications as data conversion and sampling, where precision and speed are of great importance.

You may also see an op amp refered to as being *chopper stabilized*. In such devices, the input voltage is rapidly turned on and off, or "chopped," to produce an ac waveform. This ac signal is then amplified and converted back to a dc signal at the output. This minimizes the effects of drift but results in the device being more complex and expensive than the more conventional op amp ICs.

Special *high-voltage op amps* have also been developed. The first of these, the MC1436 by Motorola, had a power supply range of from 10 to 80 V and was capable of handling input signals up to 80 V. Subsequent high-power op amps have been developed with maximum supply and input signal ratings of up to 300 V.

As mentioned earlier, the best source of information about a particular op amp is that device's data sheet plus any available applications notes. Fortunately, the same basic operating principles discussed at the beginning of this chapter generally hold for the more exotic varieties of op amp ICs. For most purposes, "garden variety" op amps such as the 741 and MC1458 will do fine.

chapter 3

POPULAR LINEAR INTEGRATED CIRCUITS

Op amps can be thought of as "general-purpose" linear ICs. Howev are not suitable for all linear and analog applications. In addition, so cuits that can be implemented using op amps and external component been implemented on a single silicon slice and contained in standarc packaging. Many such miscellaneous linear ICs have been developed. I chapter we look at some of the most widely used and interesting such dev

THE 555/556 TIMER DEVICES

The 555 timer was developed by Signetics and is currently offered by the and by several other manufacturers. The 556 is essentially two independe 555 devices in a single package. Figure 3–1 shows the pin connections fo both ICs. Both the 555 and 556 can operate from +4.5 V to approximatel +15 V, although typical supply voltages are +5 V and +9 V. Current consumption is low, with typical values being less than 5 mA at +5 V and less than 15 mA at +15 V. The maximum output current that both devices can "sink" (absorb) or "source" (supply) is approximately 200 mA, and the outputs can drive TTL devices directly if the supply voltage is held to +5 V. CMOS versions of the 555, known as the TLC555 and 7555, are available. In addition to their low power consumption compared to standard 555/556 devices, their outputs are compatible with CMOS as well as TTL.

The 555 and 556 were designed for precision timing applications, with the timing interval controlled by an external resistor and capacitor (RC) network. Inside the 555 are integrated comparators, a *flip-flop* circuit (discussed later in the book) for a definite "on" or "off" response, and an output stage. The timing intervals are independent of the supply voltage. The two basic operating modes are the *monostable* (one-shot), in which a single pulse is emitted, and the *astable,* in which a stream of output pulses is generated. In the astable mode, the 555/556 can be used as an oscillator. The relatively high current output of these devices allows them to drive LEDs, speakers, and meters directly.

Although the 555 and 556 are linear devices, they are often used in digital or "quasi-digital" applications. This is because their inputs and outputs are essentially square waves rather than sine or other complex waveforms. Input (*trigger*) signals are pulses, as are output signals; the waveforms vary between high (the supply voltage) or low (zero voltage or ground). These may be referred to as positive or negative pulses, respectively.

Figure 3–2 shows a monostable multivibrator built around one 555. This circuit responds to a *negative* trigger input and produces a positive output pulse in response. This is because a negative triggering signal turns off an internal transistor, shorting the capacitor labeled C to ground. This allows C to charge through the resistor labeled R, producing an output signal. When the charge across C is approximately two-thirds the supply voltage, the 555 can again short C to ground and the output pulse ends. The time required for this entire process is known as the *timing cycle.*

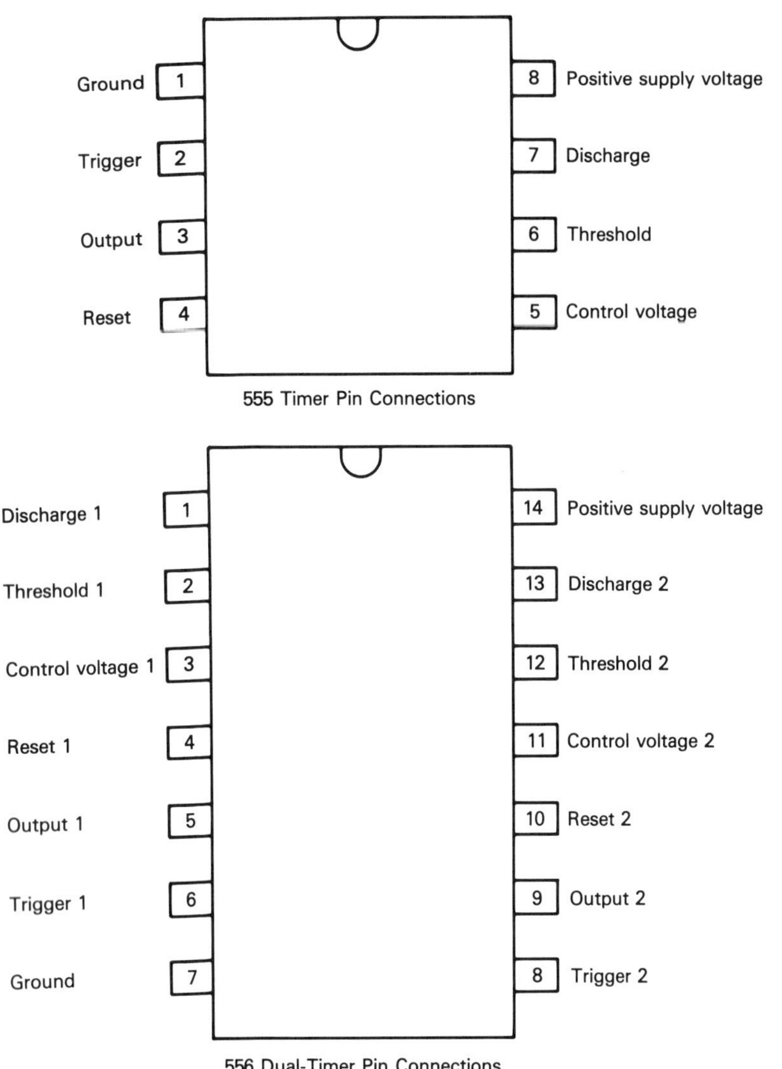

Figure 3–1 Pin connections for the 555 and 556 timers.

The output pulse is a square wave whose duration depends on the values of R and C. The formula for this is

$$\text{pulse duration} = 1.1(\text{value of R} \times \text{value of C})$$

In simple terms, the length of the output pulse increases as the values of R and C increase. For example, if C is equal to 0.1 μF and R equals 10 MΩ, the output pulse will last about 1 s. Increasing the value of C to 1 μF pro-

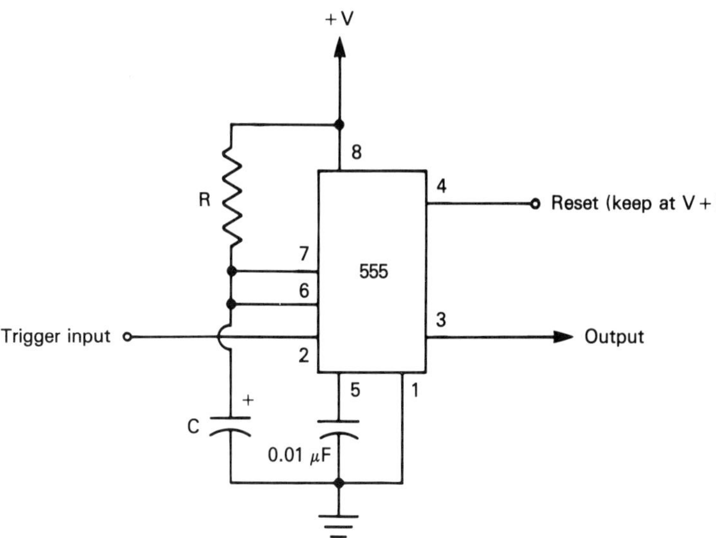

Figure 3–2 Monostable multivibrator built around the 555.

duces an output pulse duration of approximately 10 s. Longer pulses can be produced just as easily by holding the value of C constant and increasing the value of R. Once a pulse has been emitted, the circuit is "reset" by applying a "low" voltage to the reset pin (pin 4 of the 555) and the trigger signal output. Until this is done, the circuit will not respond to further trigger signals. Once reset, the reset pin should be kept at the level of the supply voltage.

Figure 3–3 shows the 555 configured as an astable multivibrator (or oscillator). Pins 2 and 6 are connected so that the circuit "triggers" itself; this is described as "free running." C charges through resistors R1 and R2 but discharges only through R2. The resulting output is a series of positive and negative pulses.

The astable multivibrator thus has two output interval times—the period when the output is high and the interval when it is low. The high output interval is found by the formula

$$\text{high output interval} = 0.693(\text{R1} + \text{R2})\text{C}$$

and the low output interval is determined by

$$\text{low output interval} = 0.693\text{R2} \times \text{C}$$

The oscillation frequency of the circuit in Figure 3–3 is found by the formula

$$\text{oscillation frequency} = \frac{1.44}{(\text{R1} + 2\text{R2})\text{C}}$$

The oscillation frequency is independent of the supply voltage.

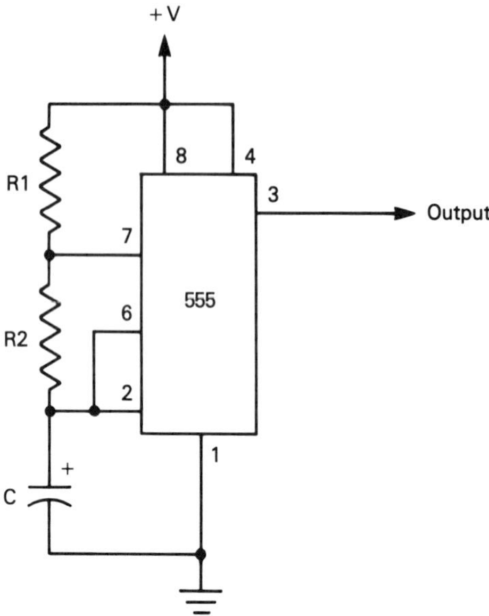

Figure 3–3 Basic astable multivibrator.

Both the monostable and astable multivibrator are subject to erratic operation due to *false triggering*. This is not always a problem; if it is, such factors as component layout may be responsible. If either circuit fails to operate properly, try grounding the control voltage pin (pin 5 on the 555 and pins 3 and 11 on the 556) through a 0.1-μF capacitor. It may also be necessary to place a capacitor in the range 0.1 to 1 μF across the ground and supply voltage pins; the proper value is determined experimentally beginning with lower values.

Although fixed-value resistors have been used in the preceding examples for simplicity's sake, they can be replaced by potentiometers. Figure 3–4 shows R1 replaced by a 1-MΩ "pot." This circuit is sometimes known as a *negative pulse generator* or *clock generator*. The stream of output pulses can be used as the "clock" signal required by many digital circuits. This can be a handy circuit to have available when experimenting with logic circuits that require such a signal. Depending on the setting of the 1-MΩ potentiometer, the output frequency can range from well over 10,000 Hz to approximately 100 Hz.

Often one needs a "burst" of square-wave output rather than a continuous stream. Figure 3–5 illustrates such a circuit. The waveform intervals and operating frequency are determined by the values of R1, R2, and C using the formula given for astable multivibrators. An output burst is caused by pressing SW, which can be any pushbutton switch. SW is connected to the supply voltage source.

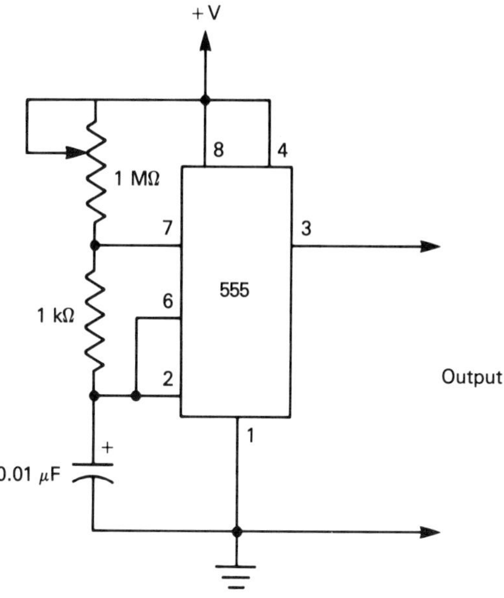

Figure 3–4 Negative pulse generator.

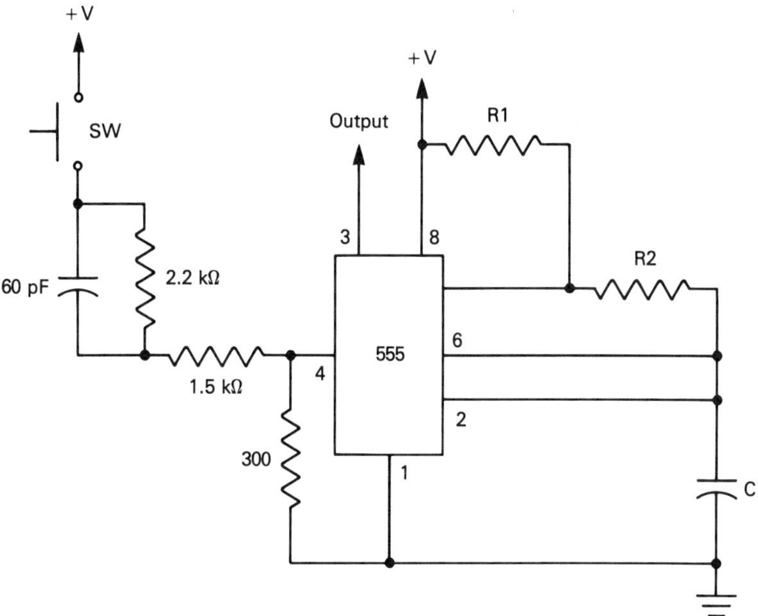

Figure 3–5 Square-wave "burst" generator.

One of the more interesting applications of the 555/556 is as a *missing pulse detector.* This circuit is a modified monostable multivibrator such that its timing cycle is synchronized with incoming pulses at its input. As long as there is an incoming pulse during each timing cycle, the output of the circuit will remain high. However, if there is not an input pulse during a timing cycle, the output of the circuit becomes low and remains so as long as the input pulses are not synchronized with the circuit's timing cycle. When a pulse does arrive "in step" with the timing cycle, the output returns to high. Thus the circuit can emit a single negative pulse if one input pulse is missing or generate a continuous "off" signal if the input pulse stream is lost or varies significantly from the timing cycle.

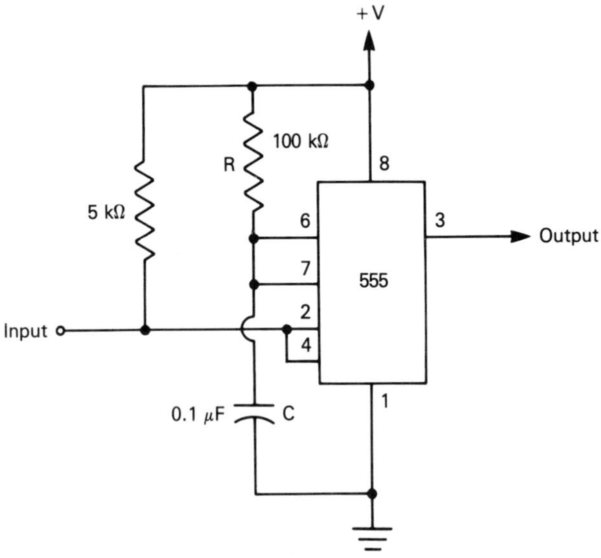

Figure 3–6 Missing pulse detector using the 555.

The frequency of the pulse detector is determined by the values of R and C in the same manner as for other monostable multivibrators; the 5 kΩ resistor remains constant. Applications of missing pulse detectors include alarms and data communications.

VOLTAGE REGULATOR DEVICES

Every circuit in this book—as well as virtually every other electronic circuit you can name—needs at least one source of stable dc voltage. It is not especially difficult to provide dc of the desired voltage from either batteries or an ac line source, but a *stable* dc voltage is another matter. The output voltage of a supply source can change with the current drawn by a load and variations in the input voltage; moreover, ac line input sources can produce "ripple" (pulsations) in the dc output voltage. To overcome these problems, *regulated* power supplies were developed many years ago using various discrete components. Although their operating principles varied, most compared the dc output voltage to a fixed reference voltage and corrected the output voltage as needed.

In 1967, Fairchild Camera and Instrument introduced the first IC voltage regulator, known as the 723. (In spite of its age, this IC is still widely used.) Additional voltage regulator ICs have been introduced since then. To power the circuits in this book, however, we only need to use one family of devices—the so-called "78XX" family. (One member of this family has already been introduced; look back at Figure 1–10 and note the 7805.)

Figure 3–7 shows a top view of the case outline for all members of the 78XX. These devices are known, for obvious reasons, as *three-terminal regulators.* The input voltage goes to pin 1, the output voltage is taken from pin 2, and the ground point is provided by pin 3; the combination mounting lug and heat sink atop the device can also be used as a ground point. There are three devices in this family. The 7805 provides a constant output of + 5 V, the 7812 a constant + 12 V, and 7815 a constant + 15 V.

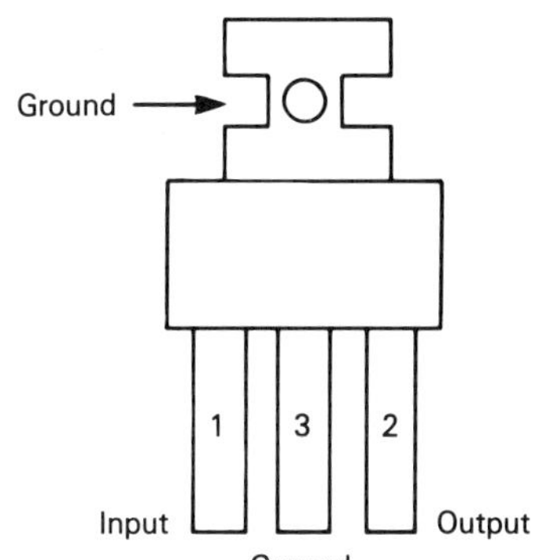

Figure 3–7 Case design and pin connections for the 78XX series of voltage regulators (top view).

All three devices can accept an input voltage of up to +35 V, with the minimum input voltage required being +2 V greater than the device's output (for a 7805, the input voltage could range from +7 to +35 V). The 78XX family can provide up to 1 A of output current. If the circuit should unexpectedly demand more (as in the case of a short circuit), the devices incorporate on-chip protection circuitry. Rather than being destroyed by excessive current demands and overheating, the IC will simply "shut down" and become inoperative. The 78XX series is also relatively inexpensive. Using these devices is simple; Figure 3–8 shows how to use a 7805 to provide a constant +5-V source for digital circuits from an input voltage of +7 to +35 V.

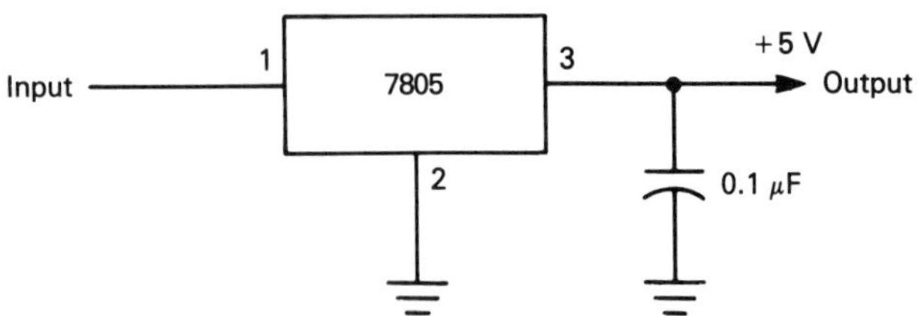

Figure 3-8 Using the 7805 for +5-V regulation.

Although the members of the 78XX family are known as "fixed" regulators, it is possible to "stretch" their output voltages and currents with some additional components. Figure 3-9 shows the configuration used to *boost* the output voltage of a 78XX device to a desired level. The output voltage obtained depends on the values of resistors R1 and R2. The values of R1 and R2 for a desired output voltage can be found by the following formulas:

$$\text{R1} = \frac{\text{normal regulator output voltage}}{0.02}$$

$$\text{R2} = \frac{\text{output voltage needed above normal}}{0.025}$$

Thus, if an output voltage of +9 V is needed from a 7805 regulator, R1 should equal 220 Ω and R2 150 Ω. If +13.8 V is desired from a 7812, R1 should be 560 Ω and R2 68 Ω. A potentiometer can be used for R2 if a variable output is desired.

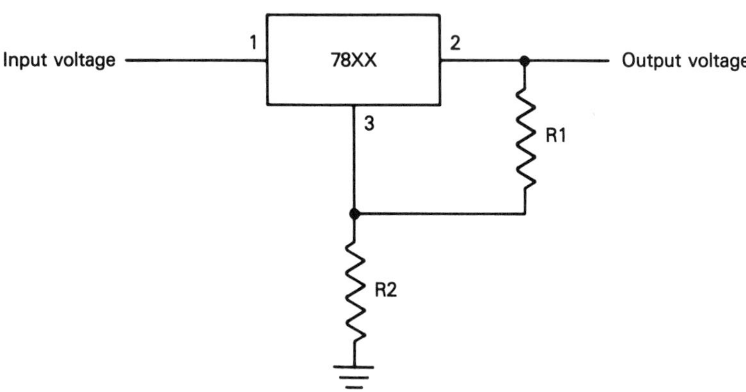

Figure 3-9 The 78XX series configured for increased voltage.

If more than 1 A of output current is required, the circuit in Figure 3-10 can be used to supply up to 4 A to the supply's load. This extra current is supplied through a PNP power transistor similar to the MJE2955 used

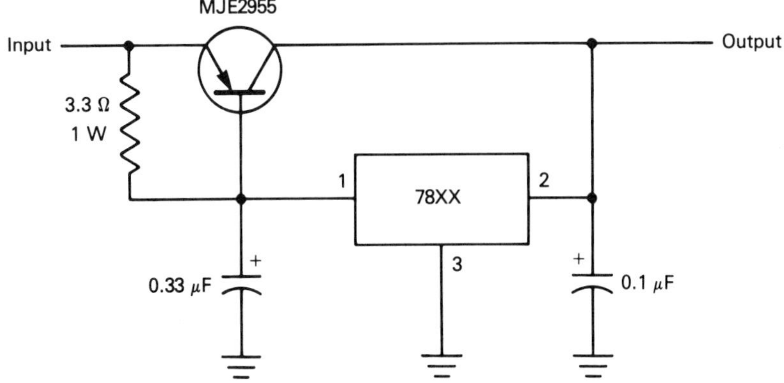

Figure 3–10 The 78XX series configured for increased current.

here; such a transistor is known as a *pass transistor.* The transistor must be mounted on a heat sink.

The 78XX series can also be used to provide a constant current as well as voltage to a load. Figure 3–11 shows a simple circuit for current regulation. The output current is determined by the output resistor, R, using the formula

$$\text{output current} = \frac{\text{output voltage}}{R}$$

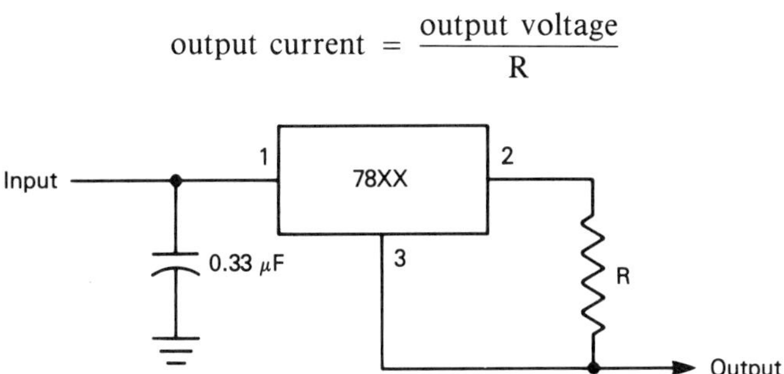

Figure 3–11 Using the 78XX series as current regulators.

One problem with power supplies providing high current is the large amounts of heat developed by the pass transistor. This can require large heat sinks and be a problem if adequate ventilation cannot be provided. The solution to this is the *switching regulator.* In this circuit, the pass transistor is rapidly switched on and off (typically, at 5 to 50 kHz). This interrupted duty cycle reduces the heat generated. The result is square-wave output which is "smoothed" by an inductor; since the switching is done so rapidly, it is easy to filter. Figure 3–12 shows a switching regulator for the 78XX series regulator. The output voltage is set by the regulator and the maximum output current is approximately 4 A.

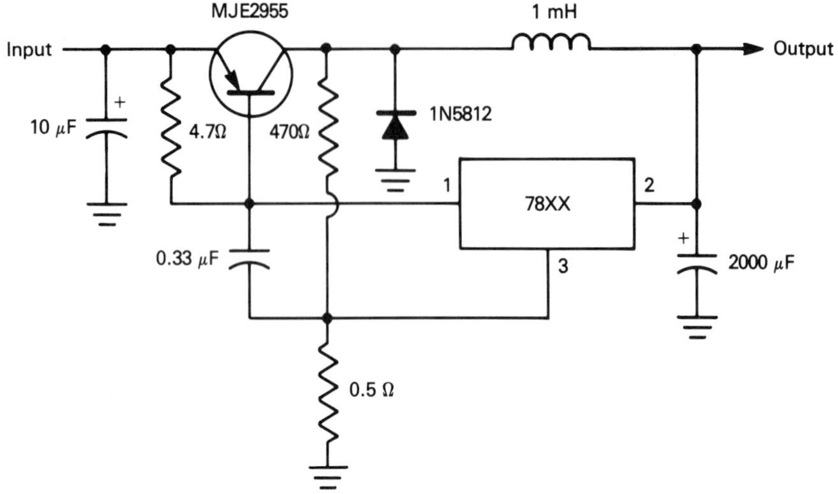

Figure 3–12 Switching regulator.

VOLTAGE-CONTROLLED OSCILLATORS

A *voltage-controlled oscillator* (VCO) in one in which the output frequency can be varied by a *control voltage* applied to it. Such circuits have been constructed for years using discrete devices, and several VCO ICs have been developed. Perhaps the most popular of these is the ICL8038 introduced by Intersil. Figure 3–13 shows the pin connections for this device.

The ICL8038 is capable of producing sine, square, triangular, sawtooth, and pulse waveforms at frequencies from approximately 0.001 Hz to over 300 kHz. Sine-, square-, and triangular-waveform outputs are available simultaneously. Frequency control can be accomplished using resistors and capacitors, and frequency modulation and sweeping are also possible. Dual- or single-polarity power supplies may be used, with maximum supply voltages of +36 V for single-polarity supplies and 18 V for dual-polarity supplies. If operated at +5 V, the outputs can drive TTL devices.

The circuit in Figure 3–14 shows a three-output audio oscillator built around the ICL8038. It can provide simultaneous sine-, square-, and triangular-waveform outputs at frequencies from 20 Hz to about 20 kHz. The output frequency is controlled by the 10-kΩ potentiometer, while the 100-kΩ potentiometer adjusts the output for minimum distortion. The 1 kΩ potentiometer controls the duty cycle of the circuit. The duty cycle can be adjusted from 2 to 98%. A dual-polarity power source is required; this circuit was designed to operate from 10 V but will operate satisfactorily from a dual-polarity 9-V source similar to that used to power many op amp cir-

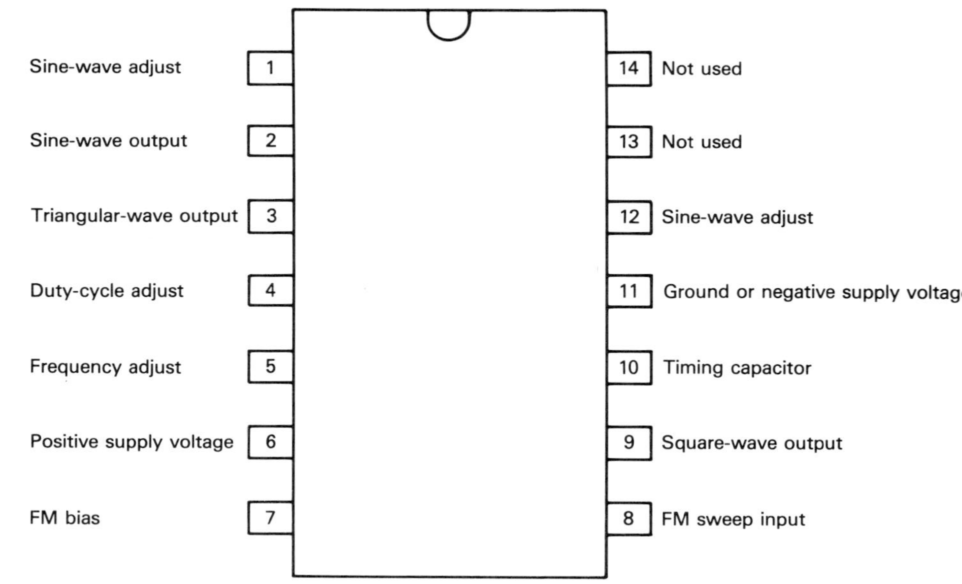

Figure 3–13 Pin connections for the ICL8038 voltage-controlled oscillator.

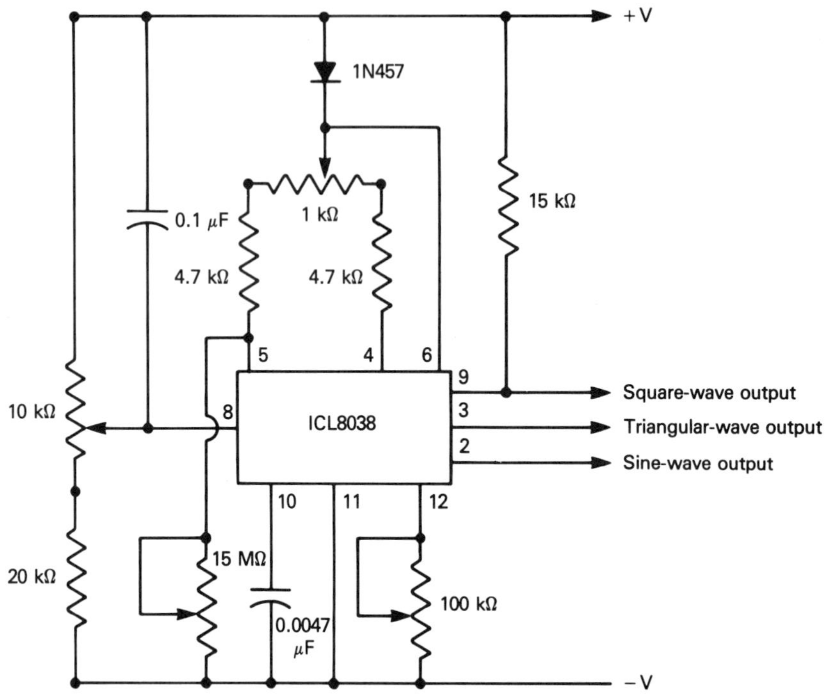

Figure 3–14 Audio oscillator with sine-, square-, and triange-waveform output.

cuits. The outputs can be used to drive speakers, audio amplifiers, and similar devices.

Most VCO circuits are not inherently linear; that is, the change in output frequency is not always proportional to the change in input voltage, and the percentage of change in the output frequency will not always be in direct relation to the percentage of change in control voltage. A linear response is preferable for many applications. To achieve this, the ICL8038 can be combined with two 741 op amps as shown in Figure 3–15. Note that the first op amp incorporates a 10-kΩ potentiometer; this is connected across the offset null terminals and is used to balance both inputs. The proper setting of this will vary, and this control must be carefully adjusted for best operation of the circuit. The 500-Ω potentiometer labeled HF is used to adjust for proper operation at high frequencies, and the 100 kΩ potentiometer labeled LF does the same for lower frequencies. The remaining 100-kΩ potentiometer connected to pin 12 of the ICL8038 is used to correct distortion in the sine-wave output. A square-wave output is available at pin 9 and a triangular-wave output is available at pin 3, although they are not used here. This circuit was designed for operation from a dual-polarity 15-V supply, although

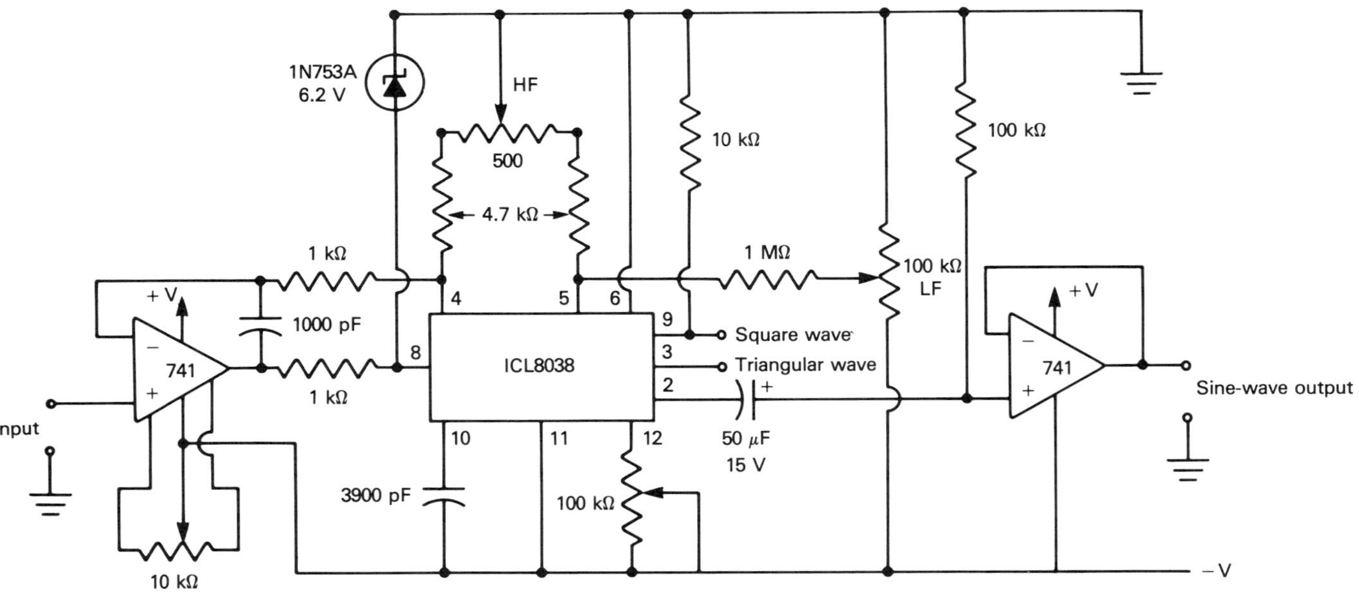

Figure 3–15 Linear voltage-controlled oscillator using the ICL8038.

a dual-polarity 9-V supply will also do. Proper adjustment of the various potentiometers is crucial for correct operation of this circuit.

FUNCTION GENERATOR DEVICES

Figure 2–40 showed a function generator circuit built from four op amps (using two MC1458 devices). This circuit is rather complex, and you may suspect that it would be possible to contain an entire function generator circuit on a single IC. A few companies have developed and produced such devices, the most popular being the XR2206 function generator IC from Exar Corporation. The pin connections for this device are shown in Figure 3–16.

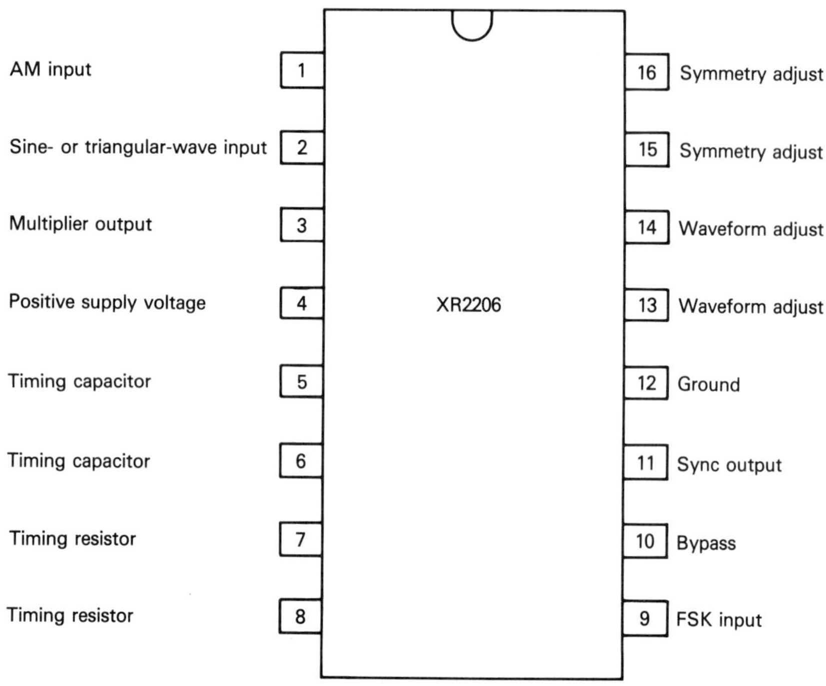

Figure 3–16 Pin connections for the XR2206 function generator IC.

The XR2206 is capable of producing sine, square, triangle, pulse, and sawtooth waveforms. Within the IC package are an op amp, a VCO, a frequency multiplier, and output stages. The output waveforms may be amplitude or frequency modulated by external voltages. The operating frequency can range from 100 Hz to over 1 MHz. The power supply voltage can range from +10 to +26 V or from a dual-polarity supply of 5 to 13 V.

Figure 3–17 shows a generator circuit capable of producing sine-,

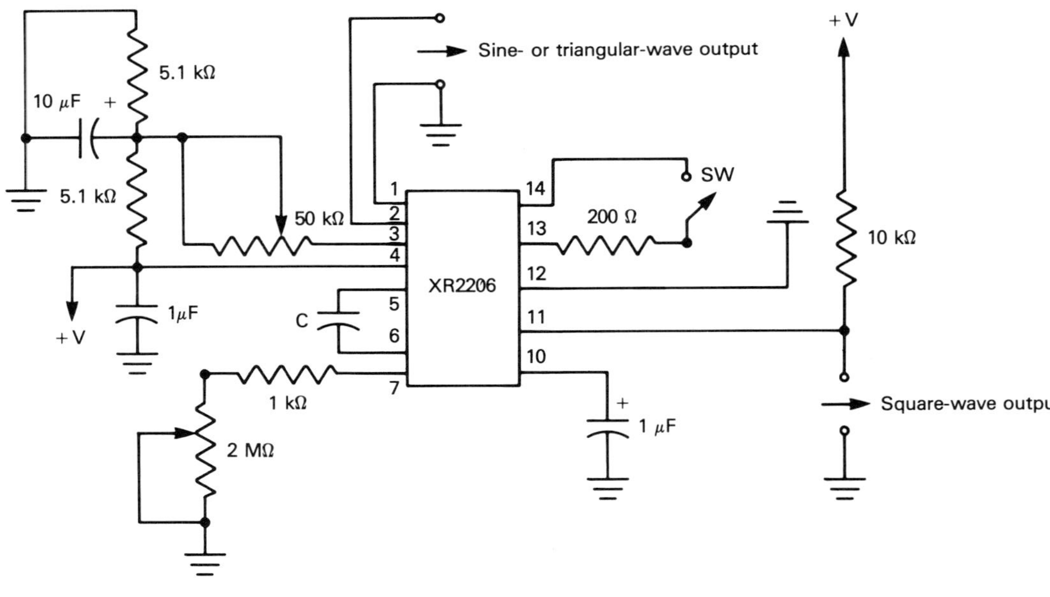

Figure 3–17 Sine-wave generator with triangular- and square-wave output.

triangular-, and square-waveform outputs. The switch SW selects between triangular- or sine-waveform outputs; if SW is open, the output is triangular, whereas the output is sine if SW is closed. If only one of those outputs is desired, the connection between pins 13 and 14 can be omitted (for triangular output) or permanently connected through a 200-Ω resistor. The amplitude of the triangular waveform is approximately twice that of the sine waveform.

The operating frequency of the circuit is determined by the capacitor labeled C in conjunction with the 2-MΩ potentiometer connected to pin 7 (it could be connected to pin 8 if desired). A resistor attached to pin 7 or 8 is known as a *timing resistor.* The optimum value of C is between 1000 pF and 100 μF; the proper value is determined experimentally by substitution. The output frequency is controlled by the setting of the 2-MΩ potentiometer. The 50-kΩ potentiometer connected to pin 3 is used to determine the amplitude of the outputs. The operating frequency is determined by the formula

$$\text{output frequency} = \frac{1}{\text{timing resistor} \times \text{C}}$$

Fixed-frequency versions of the circuit in Figure 3–17 can be constructed by substituting a fixed resistor for the potentiometer. If operation from a dual-polarity supply is desired, all ground connections can be replaced by connections to the negative-polarity supply.

Pulse and sawtooth waveforms can be produced by the circuit in Figure 3–18. The operating frequency (pulse width) and duty cycle are set by resis-

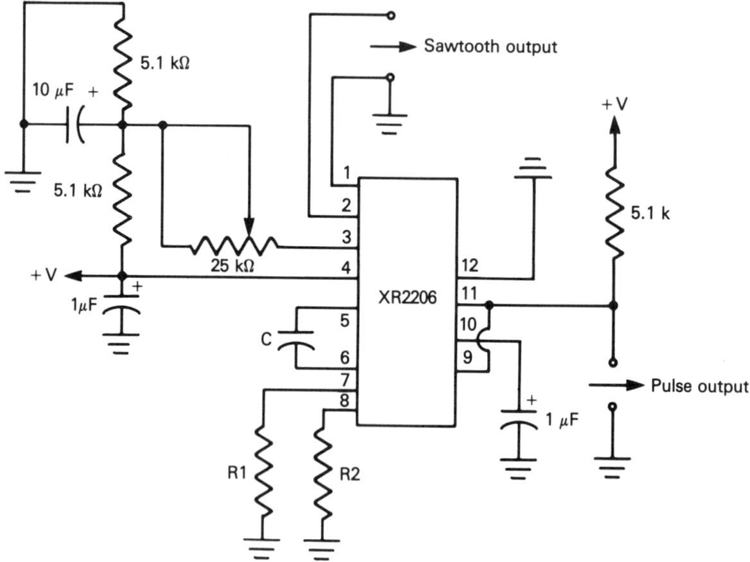

Figure 3–18 Pulse and sawtooth waveform generator.

tors R1 and R2 together with capacitor C. The pulse width of the circuit is determined by the formula

$$\text{operating frequency} = \frac{2}{C} \times \frac{1}{(R1 + R2)}$$

while the duty cycle is found by

$$\text{duty cycle} = \frac{R1}{R1 + R2}$$

The values for R1 and R2 should lie in the range 1 kΩ to 2 mΩ.

PHASE-LOCKED-LOOP DEVICES

One of the most useful circuits in contemporary electronics is the *phase-locked loop* (PLL). Figure 3–19 illustrates a block diagram of the operating principle of a PLL. The heart of a PLL is a VCO circuit whose output is sampled by a *phase detector*. The phase of the VCO's output signal is compared to the phase of the signal generated by the *reference oscillator*. If the phase differs, the phase detector generates a *correcting voltage*. This correcting voltage is sent through a low-pass filter, which determines the *capture range* of the loop circuit. This is the frequency range over which the PLL "locks" on the desired output frequency. After the low-pass filter, the correcting voltage is amplified and then applied to the VCO. In this way, the frequency of the VCO can be stabilized at its desired value to a remarkable degree.

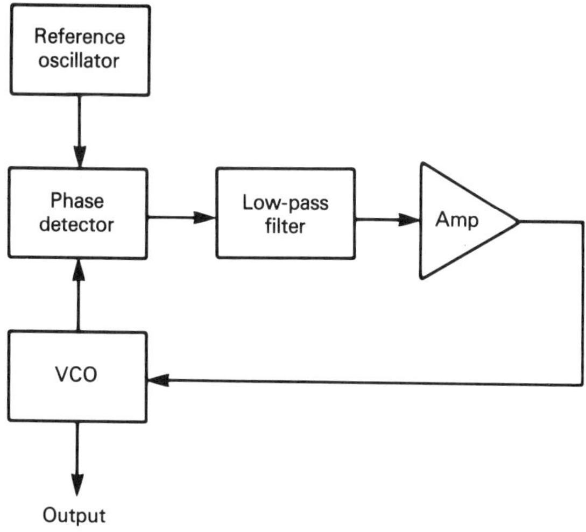

Figure 3–19 Block diagram of a phase-locked loop.

PLL circuits are remarkably versatile and are used in such devices as radio receivers, modems, tone decoders, FM detectors and discriminators, frequency multipliers and dividers, frequency-shift keyers, data synchronizers, and many others. PLLs have been implemented using discrete components and in integrated form. One of the most popular PLL ICs is the LM565 from National Semiconductor. Figure 3–20 shows the pin connections for this device.

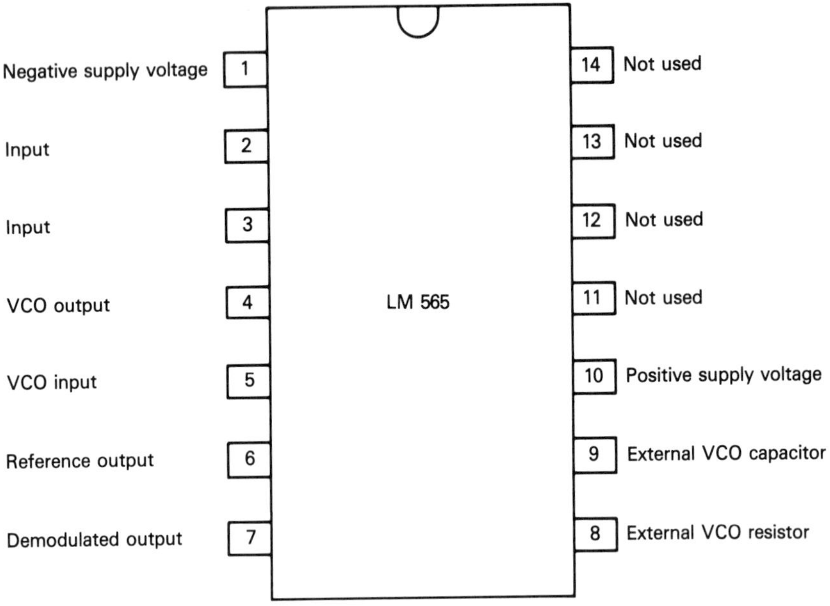

Figure 3–20 Pin connections for LM565 PLL IC.

The LM565 is capable of operating at an output frequency of 10 Hz to 500 kHz. It usually operates from a dual-polarity power supply ranging from 6 to 12 V, although a single-polarity supply of twice these voltages can be used. The maximum input voltage rating is 3 V. The operating frequency can be determined by a resistor or capacitor as well as by voltage or current values. If operated for square-wave output, it can directly drive TTL circuits. The output is highly linear.

The major applications of the LM565 is in communications circuits, such as FM receivers, modems, telemetry receivers, and tone decoders. One example is the circuit in Figure 3–21 for a *subsidiary communications authorization* (SCA) *decoder*. Many FM broadcasting stations in the United States have SCA programming; this is typically background music, features for professional groups such as doctors and lawyers, "talking books" for the visually

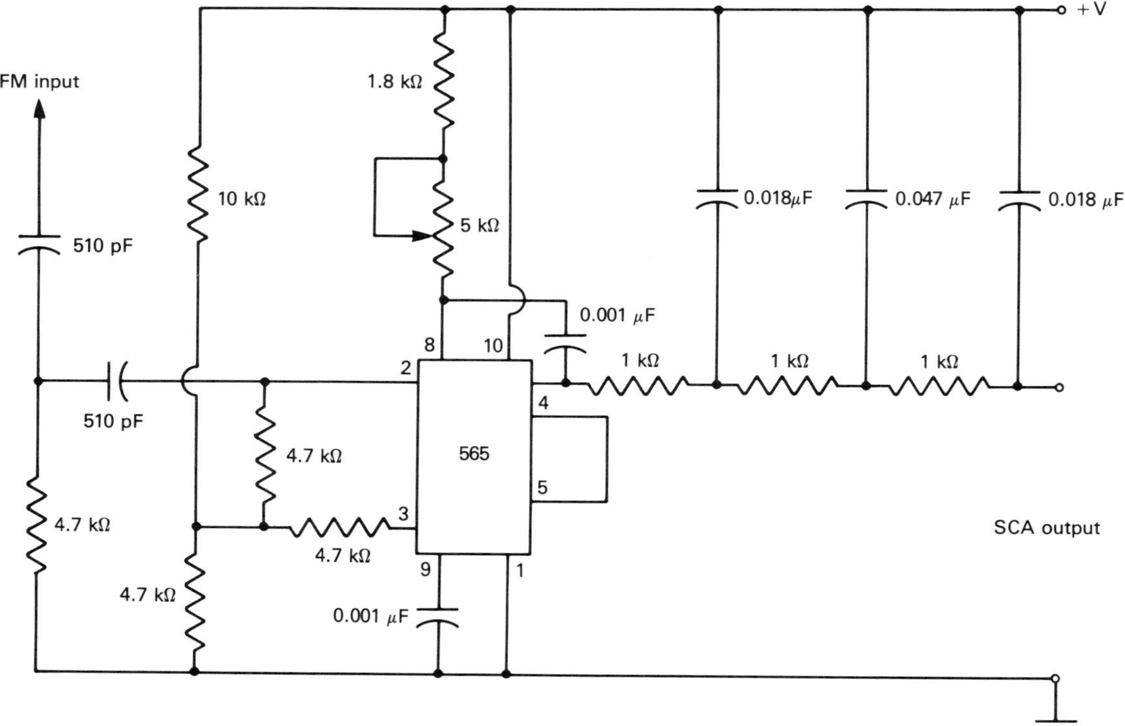

Figure 3–21 SCA decoder using the LM565.

handicapped, special time signals, and other materials of interest to specific narrow segments of the population. SCA programming is transmitted simultaneously with normal station programming, using an FM subcarrier frequency of 67 kHz and reduced amplitude, so as not to interfere with normal programming. Receiving this SCA subcarrier requires a SCA decoder.

The circuit in Figure 3–21 will filter out and demodulate any 67-kHz SCA signal present in the signal from a FM broadcast station. The PLL is tuned to 67 kHz by adjusting the 5-kΩ potentiometer connected to pin 8. Only rough adjustment of this potentiometer is necessary, since the loop circuitry will "seek" the 67-kHz SCA signal and "lock" onto it. The input signal should have an amplitude between 80 and 300 mV and be obtained from an input impedance of less than 10 kΩ. Note that a single-polarity supply is used and the supply voltage can be between +12 and +24 V. The demodulated SCA output is approximately 50 mV and has a maximum frequency response of approximately 7 kHz. (Since much SCA material is either spoken voice or intended for inobtrusive background listening, this is usually not a significant drawback.)

Many other PLL ICs have been developed for specific purposes, particular for use in modems, frequency-shift keying, and tone decoding.

LM3909 LED FLASHER AND OSCILLATOR

A favorite IC with many experimenters and hobbyists is National Semiconductor's LM3909. This is a monolithic IC oscillator designed specifically for flashing light-emitting diodes in low-power applications. The rate at which the oscillator operates is controlled by external timing capacitors and resistors. Figure 3–22 shows the pin connections for this device.

The LM3909 works by boosting a supply voltage of 1.5 V to over 2 V in short pulses. It has been designed for minimal battery drain; a single

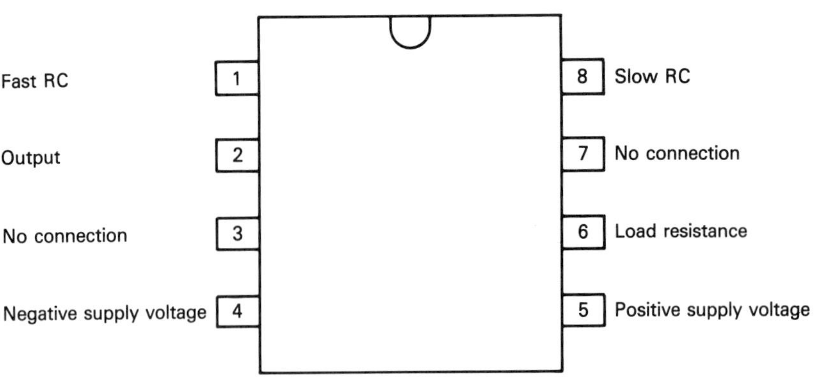

Figure 3–22 LM3909 LED flasher/oscillator-device pin connections.

1.5-V C cell can flash a LED for over a year in continuous service. Figure 3–23 shows just such a circuit. The rate at which the LED is flashed depends on the value of the capacitor connected between pins 1 and 2. In this circuit, the flash rate will be approximately 1 Hz (or one flash per second). Electrolytic capacitors are used for timing. The flash rate can be slowed by placing resistance in series with the timing capacitor and a variable flash rate can be obtained by using a potentiometer in place of a fixed resistor.

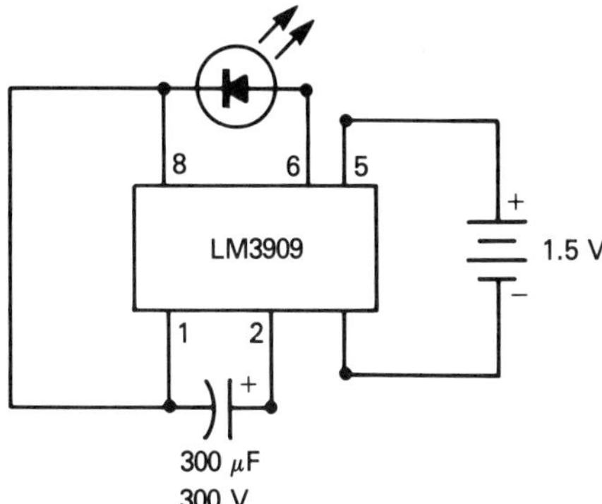

Figure 3–23 1.5-V LED flasher.

The output of the LM3909 is sufficent to drive a small 8-Ω speaker, and the IC can be used to produce audio tones as well as to flash a LED. The circuit in Figure 3–24 operates from a 3-V supply and produces a tone in the speaker when the switch is closed. If a telegraph key is substituted for the switch, this circuit can be used as a Morse code practice set.

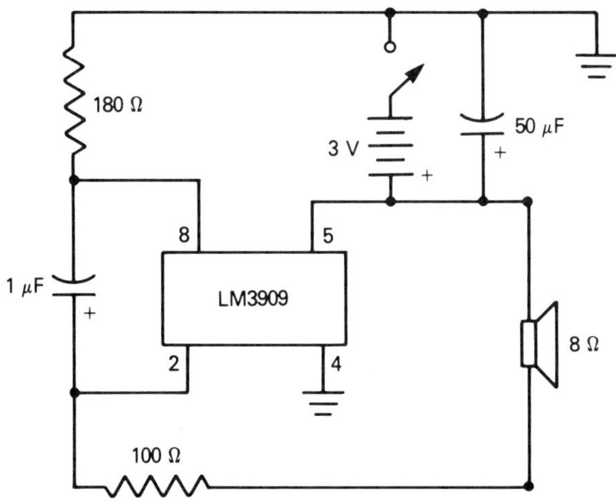

Figure 3–24 Tone oscillator using the LM3909.

chapter 4

TTL DIGITAL CIRCUITS

Back in Chapter 1, we discussed briefly the history and development of digital ICs. (You might find it helpful to look back at that section now, particularly the schematic symbols in Figure 1-6.) Although the RTL (resistor-transistor logic) and DTL (diode-transistor logic) families came first, it was the introduction of TTL (transistor-transistor logic) devices, pioneered by Texas Instruments, that really started the widespread use of digital logic ICs in all areas of electronics in the early 1970s. TTL soon became the most widely used logic family in the world. Today, it has fallen behind the CMOS device family in popularity and has some marked technical disadvantages compared to CMOS. However, TTL is still a good choice for many applications and will probably continue to be widely used for many years to come.

The popularity of TTL was no accident. It was (and is) inexpensive. It is also a "fast" logic technology, offers fair noise immunity, and its fan-in and fan-out capabilities are adequate for many applications. It is also a "rugged" device family, having good immunity from damage due to random static electricity discharges. The popularity of TTL technology resulted in numerous TTL devices being produced; Table 4–1 shows some of the more popular devices and their component numbers. The sheer number of TTL devices makes it impossible to cover them in all this chapter, however.

TABLE 4-1 Major TTL IC Devices

Number	Description
7400	Quad two-input NAND gate
7401	Quad two-input NAND gate, open collector
7402	Quad two-input NOR gate
7404	Hex inverter
7405	Hex inverter, open collector
7408	Quad two-input AND gate
7410	Triple three-input NAND gate
7411	Triple three-input AND gate
7413	Dual four-input Schmitt trigger
7414	Hex Schmitt trigger
7420	Dual four-input NAND gate
7421	Dual four-input AND gate
7427	Triple three-input NOR gate
7430	Eight-input NAND gate
7432	Quad two-input OR gate
7437	Quad two-input NAND buffer
7441	BCD-to-decimal decoder
7447	BCD-to-seven-segment decoder/driver
7448	BCD-to-seven-segment decoder/driver
7451	Dual AND-OR-Invert gate
7453	Expandable four-wide AND-OR-Invert gate
7454	Four-wide AND-OR-Invert gate
7473	Dual J-K flip-flop with clear
7474	Dual D-Type positive-edge-triggered flip-flop with reset and clear
7475	Four-bit bistable latch

TABLE 4-1 (Continued)

Number	Description
7476	Dual J-K flip-flop with preset and clear
7485	Four-bit magnitude comparator
7486	Quad XOR gate
7490	Decade counter
7491	Eight-bit shift register
7492	Divide-by-12 counter
7493	Divide-by-16 counter
7495	Four-bit left/right shift register
74109	Dual JK positive-edge-triggered flip-flop
74121	Monostable multivibrator
74123	Dual retriggerable monostable multi-vibrator with clear
74132	Quad two-input NAND Schmitt trigger
74138	1-of-8 decoder multiplexer
74139	Dual 1-of-4 decoder
74151	1-of-8 data selector
74153	Dual four-input multiplexer
74154	Four-line-to-16-line decoder
74157	Quad two-input multiplexer
74161	Four-bit up counter
74164	Eight-bit shift register
74165	Eight-bit parallel-to-serial converter
74173	Four-bit D-type register
74174	Hex D-type flip-flop
74175	Quad D-type flip-flop
74190	Up/down decade counter
74191	Up/down binary counter
74192	Up/down decade counter with up/down clocks
74193	Up/down binary computer with up/down clocks
74194	Four-bit shift register
74196	Presettable decade counter
74221	Dual monostable multivibrator
74240	Octal buffer/line driver with three-state output
74241	Octal buffer/line driver with three-state output
74244	Octal buffer/line driver with three-state output
74245	Octal bus transceiver with three-state output
74367	Hex three-state buffer
74368	Hex three-state bus driver
74373	Octal transparent latch with three-state output
74374	Octal D-type flip-flop with three-state output

The major disadvantage of TTL has always been its heavy current consumption. Although versions of TTL consuming less current have been developed, they are still usually inferior to CMOS in this respect. Moreover, the comparatively large current demanded by TTL can introduce "glitches" (or variations) into the power supply, which can cause erratic operation in other TTL circuits operating from the same supply. In fact, false "triggering" of TTL gates due to power supply variations may be the most common problem in building and troubleshooting TTL circuits. Layout of the components in a circuit using TTL and the length of leads between them can be critical, especially in circuits where logic states switch rapidly. TTL also requires a constant +5-V power source (referred to as *Vcc* in this chapter) and a TTL device can be destroyed quickly if this level is exceeded.

Although CMOS is heavily favored over TTL in contemporary design work, TTL makes a good choice for the experimenter or hobbyist. Its low cost and "sturdiness" make it ideal for those who do not have a fortune to spend on replacement ICs, and the wide variety of devices available offer many design and experimentation options.

The number of TTL devices available and their applications are so numerous that entire books could be—and have been—written on them. As mentioned previously, in this chapter we examine some of the most popular TTL devices and applications; however, any "logical" digital circuit (i.e., one in which the outputs are the product of the inputs and rules of the logic functions) can be implemented using TTL. This means that TTL is excellent for learning about digital circuits as well as practical implementations of them. As long as the operating requirements are met and proper pin connections are observed, the only penalty in most cases for an improper design will be a nonfunctioning circuit.

"FLAVORS" OF TTL DEVICES

TTL devices can be easily identified by their part numbers, which normally consist of four or five digits and begin with the digit "7." (Some early TTL devices used similar part numbers beginning with "5," and these are sometimes encountered today.) A few TTL devices have part numbers beginning with "8," but most of these also have an equivalent number, usually five digits long, beginning with "7."

Letter codes are often found inserted into TTL part numbers after the first two digits. For example, in addition to the "plain vanilla" 7400 TTL IC, you will also run across such variations as 74ALS00, 74C00, 74LS00, and others. These letters indicate the "subfamily" of TTL to which the device belongs (although in some cases the device is not really TTL). In terms of pin connections and functions, these devices are identical to the ordinary TTL equivalent. However, they are *operationally* different; they vary in such parameters as power consumption and speed and are used in place of standard TTL, depending on the application.

What do these letter codes mean? Following is a brief summary of what they tell you about the devices.

ALS. This stands for *advanced low-power Schottky,* a technology that significantly enhances the speed and lowers the power consumption of TTL devices. This is the "highest-performance" TTL subfamily overall, comparing favorably with CMOS in many respects. It has become a design favorite of many professional engineers as a result. However, for most hobby and experimental applications, it is "overkill" for the price. If low power consumption is important, CMOS is usually the more cost-effective solution for most situations.

C. WARNING! *This indicates the CMOS equivalent of a TTL device!* Do not let the TTL part number fool you—although the pin connections and device functions are the same as those for the comparable TTL device, a chip like the 74C00 is a CMOS device through and through. It must be used and operated according to the limitations of CMOS, and failure to take these into account (such as protection from static discharge) can ruin the device. It also means that special interfacing is necessary if TTL and CMOS devices are combined in a single circuit.

F. This stands for *fast,* and is a version TLL optimized for maximum speed.

H. This represents *high power,* an early TTL subfamily that offered higher speed at the price of greater power consumption. It is seldom used today.

HC. ANOTHER WARNING! This indicates the *high-speed CMOS* equivalent of a TTL IC. The same comments for the "C" letter code also apply here.

L. This stands for *low power,* and was an early attempt to reduce power consumption by sacrificing speed. It is seldom used today.

LS. This indicates a *low-power Schottky* TTL IC. This is a very successful and popular subfamily that features greatly reduced power consumption (on the average, about 20% of standard TTL) and somewhat increased operating speed. However, it is more expensive than ordinary TTL. For many applications, this represents a good compromise between performance and cost.

S. As you might suspect, this indicates a *Schottky* TTL device. It was named for the fact that Schottky diodes were incorporated into the transistor-transistor logic gates integrated onto the silicon inside the package. The result was improved speed. This is another subfamily that is seldom used today.

Given this wide selection, deciding on which version of TTL to use can be confusing. For the purposes of building the circuits in this book, the only practical choices are "plain" TTL and the LS variety. Ordinary TTL has been used for the circuits in this book and is a good choice for experimenting or learning about TTL. Moreover, if you decide to build a permanent version of a circuit that is to be powered from an ac line supply, the high power consumption of TTL is no longer a significant factor. If a circuit is to be powered from batteries, LS is the best choice. If a circuit's speed is important (as in a counter or divider), LS is also preferable. The CMOS equivalents (C and HC codes) can be used if the circuit is to be used with other CMOS devices and circuits, since it simplifies power supply considerations and eliminates the need for interfacing between CMOS and ordinary TTL devices.

You will sometimes see some TTL devices described as *open collector.* There are some open-collector equivalents to popular TTL devices; for example, the 7401 is identical to the 7400 except that the 7401 is an open-collector device. The difference between the two is that the 7400 has output transistors integrated into it to provide a low output impedance regardless of whether the output is high or low, while the 7401 does not have such transistors integrated. When using open-collector devices, an external resistor, known as a *pull-up resistor,* must be used between the output and any other device, component, or input. The advantage of open-collector devices is that several different outputs can be "wired" together to form a single output. The major use of open-collector devices is in connecting logic outputs to a *data bus* in microcomputer and data communications systems.

USING TTL

For proper operation, there are several operating considerations that must be observed. Failure to heed these can result in improper circuit operation or even destroy the TTL device.

The supply voltage must be held as close to +5 V as possible; in no case should the supply voltage exceed 5.25 V. The supply voltage should also not be less than 4.75 V. If an ac line power supply is used, a regulated power supply (such as the one in Figure 1–10 using a 7805 voltage regulator IC) is essential. The leads from the power supply to the TTL devices should be as short as possible. A capacitor in the range 1 to 10 μF should be across the power supply leads where they make contact with a solderless breadboard.

To prevent improper operation due to power supply variations (noise) as the gates switch states, "deglitching" capacitors (also known as *bypass* or *decoupling* capacitors) should be used when a circuit uses more than one TTL device. A good rule of thumb is to place a 0.1-μF capacitor across the Vcc and ground pins of each TTL device in circuits using two or more TTL devices. The leads of such capacitors should be kept in short as possible. Higher-speed TTL varieties, such as LS, are more susceptible to noise than is standard TTL.

Most TTL devices will treat any input signal in excess of +2 V as a high input signal and any signal of +0.8 V or less as a low input signal. For best results, input signal levels should comfortably exceed these minimum levels; preferably, a high input signal should equal Vcc and a low input signal should be at ground (0 V). Input voltages must not exceed Vcc. All unused inputs on a device should "go somewhere" and not be left floating; if an input is not being used to receive a signal that switches states, it must be held high (connected to Vcc). It is possible to apply to the same input signal to two or more inputs by connecting all inputs to a common point to which the input signal is applied. (Several examples of this are presented later in the chapter.) If a battery power supply is used or if power consumption is crucial, the outputs of any unused gates should be set to high by connecting the gate's inputs to Vcc or ground as appropriate.

Fan-out capabilities vary with the version of TTL used. A good rule of thumb is that each TTL output can "drive" no more than 10 TTL inputs or 20 LS inputs. A LS output can drive half the number of inputs TTL can (5 TLL inputs or 10 LS inputs). If a device is operated at these fan-out limits and the circuit containing it operates improperly, try reducing the number of inputs the outputs drive.

COMBINATIONAL AND SEQUENTIAL LOGIC

There are two terms used to describe the logic circuits constructed from TTL and all other digital IC families. *Combinational* logic circuits respond based on their inputs at any given moment regardless of any prior state or condition the circuit may have been in. By contrast, the operation of *sequential* logic circuits does depend on the prior state or condition of the circuit. Whereas a combinational circuit can be thought of as reacting "instantly" to an input signal, a sequential circuit seems to be operating in increments or steps. (Given the speed with which a digital circuit can operate, however, a sequential circuit can appear to be operating "instantly" even though it is incremental in nature.) A sequential circuit in operation can be thought of as a line of stacked dominos in which the first one is tipped over; each change of state is part of a chain in which each device (except the first) is affected by prior devices and in turn affects each subsequent device (except for the last device).

Neither term refers to how simple or complex a logic circuit is. Both combinational and sequential circuits may be uncomplicated or very elaborate.

You may have anticipated one problem with sequential logic. In complex sequential circuits composed of several components, the desired result or operation may depend on several different devices operating in proper sequence. To coordinate the operation of these devices, a signal, known variously as a *clock, enable,* or *trigger,* is often provided for them. Without this signal, the devices will not emit certain outputs or respond to input signals. A component or circuit whose operation is governed by a clock signal is called *synchronous.* Circuits or devices that do not require clock signals are said to be *asynchronous.* Several such devices and circuits are examined in this chapter and the next.

THE 7400 QUAD-NAND-GATE DEVICE

The "universal" TTL IC is the 7400, which includes four independent two-input NAND gates in a single package. This is an extremely popular device because, as mentioned in Chapter 1, *any* logic gate can be implemented using a network of NAND gates. Figure 4–1 shows the pin connections and internal logic of the 7400.

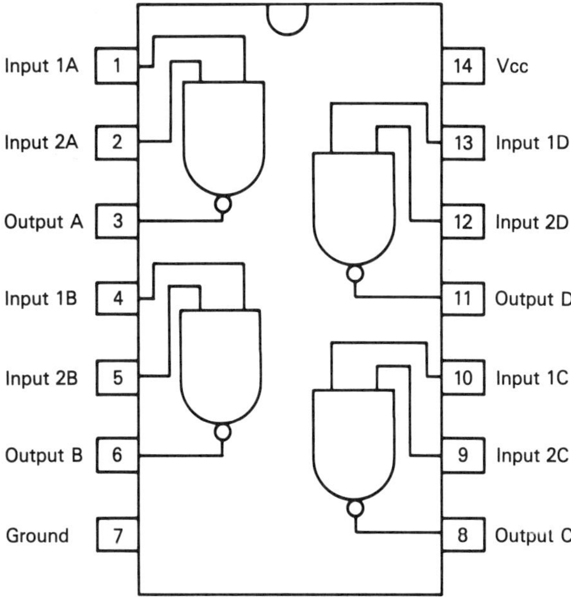

Figure 4–1 Pin connections and internal logic for 7400 quad-NAND-gate device.

Figure 4–2 shows how to construct other logic gates using the NAND gates on a single 7400 device. (These are all combinational logic circuits.)

The inverter circuit is implemented simply by connecting both inputs together and applying the same input signal to both; the output is the opposite of the input. The inverter is used in all the other logic gates in Figure 4–2. The AND gate is nothing more than an ordinary NAND gate followed by an inverter. As with all AND gates, a high signal is necessary at both inputs to produce a high output.

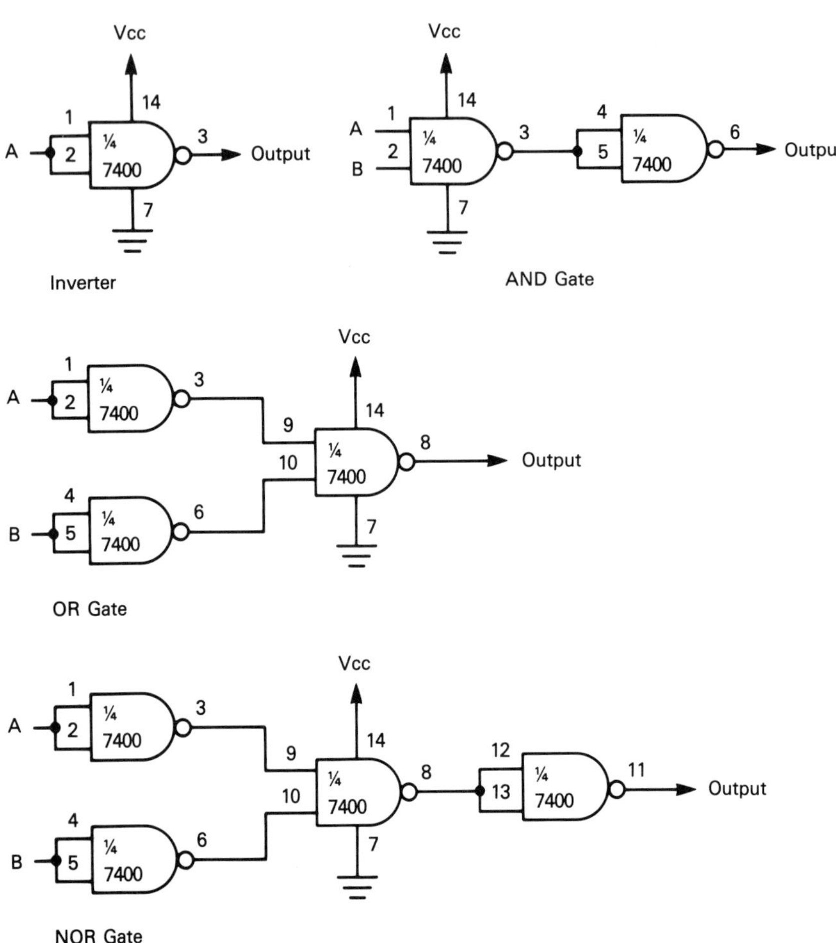

Figure 4–2 Basic logic gates using the 7400.

The OR gate uses an inverter at each input; their outputs serve as the inputs for a standard NAND gate. A high input signal at either A or B, or both, will produce a high output signal. Three inverters are used to implement a NOR gate, which is an OR gate followed by an inverter. The action of the NOR gate is therefore the opposite of the OR gate; its output will be high only if both inputs are low. A high input signal at either or both will "turn off" the NOR gate output.

Two more complex NAND networks are shown in Figure 4-3. "XOR" is an acronym for the *exclusive-OR* gate. An XOR gate produces a high output if the input signals differ, but the output is low when the input signals are the same. That is, the output is low when the input signals are either both low or both high; the output is high when one input is low while the other input is high. The XOR gate in Figure 4-3 is an interesting design, as

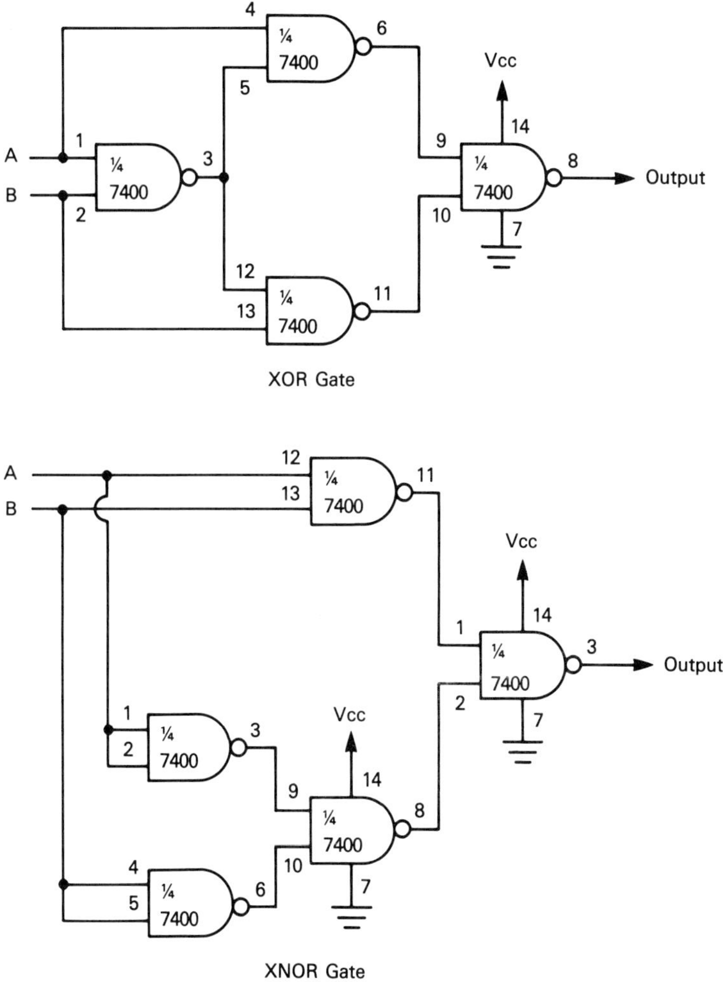

Figure 4-3 XOR gate and XNOR gate using the 7400.

each input signal is applied to two different gates and the output of the first NAND gate is used as one input for two NAND gates that use the original input signals as the other input.

The XNOR (*exclusive NOR*) gate shown in Figure 4–3 requires five NAND gates and thus must be implemented using more than one 7400 device. The action of the XNOR is the opposite of the XOR gate; if the inputs are identical (both low or both high), the output is high. If the input signals differ, the output is low. Although implementing an XNOR gate using NAND gates is somewhat complex, it does illustrate how it is indeed possible to produce any of the basic logic gates using only NAND gates. This is why several 7400 devices are useful to have around, since in a pinch they can be substituted for other gates in circuit design and then replaced by more appropriate devices, such as the 7486 quad-XOR-gate device.

One of the more useful digital circuits is the *flip-flop.* This is sometimes known as a *bistable multivibrator.* The output of this circuit is either a low- or high-level logic signal. The output stays low or high until an input signal known as the *trigger* is received; when it is, the output changes to the opposite state. This is the simplest form of a flip-flop circuit, and most "real-world" flip-flops are more complex in their operation. (Since their output depends on prior states or conditions, flip-flops are sequential, rather than combinational logic circuits.)

Flip-flop circuits are available in IC form, and these will be examined later. It is possible to implement flip-flops using NAND gates, however, and this is an easier way to understand their principles of operation. Figure 4–4 shows one type of flip-flop known as an *R-S (reset-set) flip-flop.* The circuit is also known as a *latch.* This circuit has two outputs, Q and $\overline{Q}$. Both Q and $\overline{Q}$ can be either high or low, but they must always be *opposite* to each other. Such outputs are said to be *complementary.* (This description comes from the mathematical concept of the complement of a number.) The states of Q and $\overline{Q}$ depends on the signals present at the *set* input (S) and the *reset* input (R). The operation of this circuit can be illustrated by the following logic tables:

S	R	Q	$\overline{Q}$
L	L	Not allowed	
L	H	H	L
H	L	L	H
H	H	No change from previous state	

In the table the term "not allowed" refers to a condition where Q and $\overline{Q}$ would have the same logic level, which is not permitted by the definition a flip-flop circuit. When the outputs of a flip-flop circuit change (switch "states") in response to changes in its input signals, the circuit is said to *toggle.*

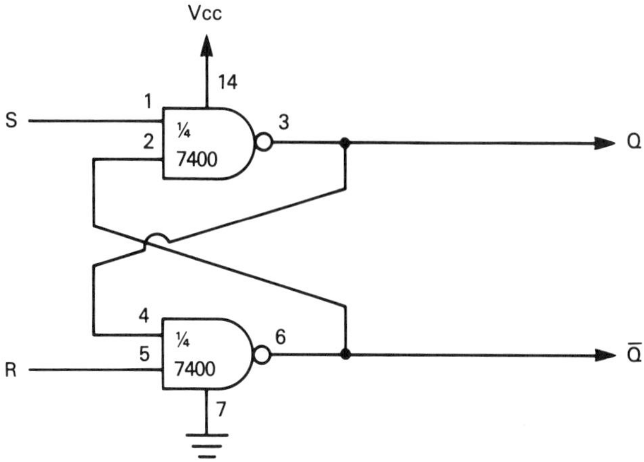

Figure 4–4 R-S flip-flop using the 7400.

The R-S flip-flop action in Figure 4–4 is controlled entirely by the signals at R and S. However, there are many situations where it is preferable to have the flip-flop's response to the signals at R and S depend on another control signal, known as the *enable* or *clock* signal. Figure 4–5 shows an *enabled* or *clocked R-S flip-flop* circuit using four NAND gates. The circuit "ignores" the signals at R and S until a pulse is received at input E. The operation of this circuit is described by the following table:

S	R	Q	$\overline{Q}$
L	L	No change from previous state	
L	H	L	H
H	L	H	L
H	H	Not allowed	

Note that this table is valid *only* when a positive pulse or signal is present at E; otherwise, the circuit maintains its previous state.

Since the R-S flip-flop does not require a clock signal for operation, it is asynchronous; the enabled R-S flip-flop does and is therefore a synchronous circuit.

Flip-flop circuits are actually the most basic examples of *memory* circuits. The outputs, Q and $\overline{Q}$ can represent one bit of data in binary form; since the ouput will not change unless new "data" are received in the form of changing input signals, the flip-flop can be said to actually "store" data. As you might expect, the enabled or clocked versions of flip-flop circuits are preferred, since this gives us even greater control over the "reading" and "storage" of data. Flip-flop circuits may seem crude compared to today's

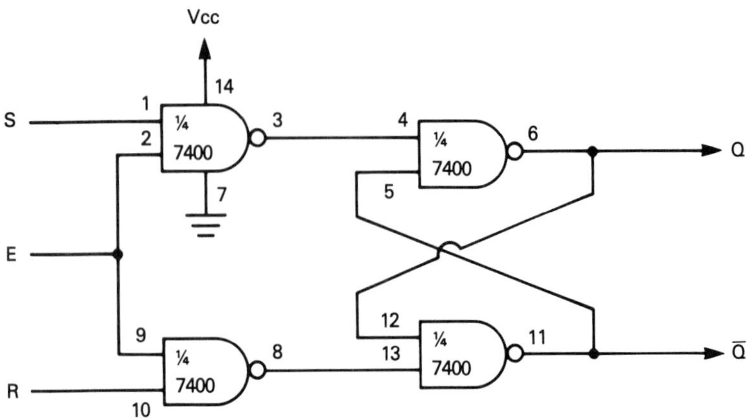

Figure 4-5 "Enabled" version of the R-S flip-flop.

high-capacity semiconductor memories, but their functions are remarkably similar. In fact, the earliest computers used numerous flip-flops, implemented using vacuum tubes, to store and manipulate data. Later in this chapter we examine some of these circuits that can be implemented using flip-flop ICs.

Often a constant high or low logic signal is useful. Figure 4-6 shows how two NAND gates can be used to produce either a high- or a low-level signal, depending on the setting of the switch as indicated. Although a manual switch has been used in Figure 4-6, several alternative switching methods, such as a relay or a pushbutton, can also be used.

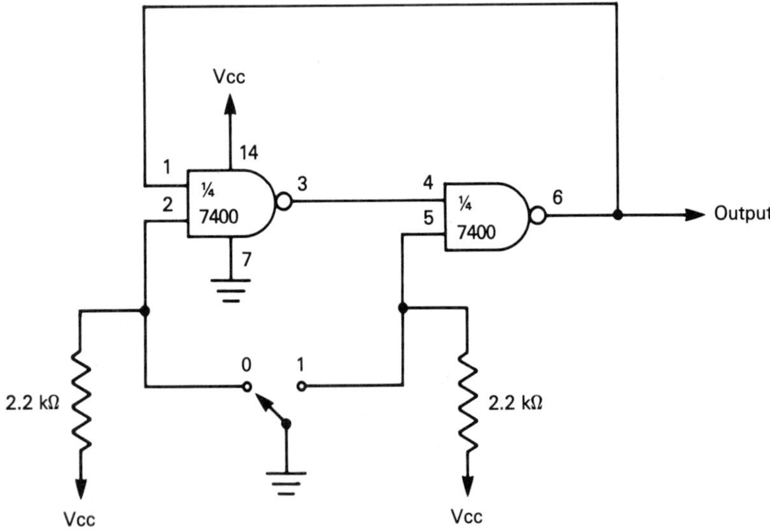

Figure 4-6 A 7400 used to generate selectable high and low outputs.

Figure 4–7 shows a pushbutton used with a NAND gate to create a *"bounceless" switch.* This name comes about from the fact that most mechanical switches used in logic circuits (such as the keyboard of a microcomputer) can generate more than one pulse when closed due to the inherent mechanical "spring" in a pushbutton switch. The NAND gate introduces a delay between the time the switch is closed and the output of a high logic signal. This means that one—and only one—high-level pulse is emitted each time the bushbutton is pressed. This circuit can be a useful tool in the design and debugging of logic circuits, since it is a reliable source of a single logic pulse.

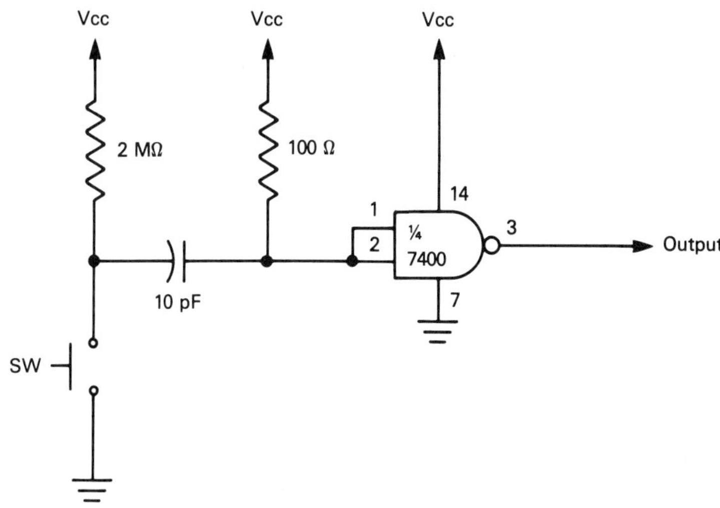

Figure 4–7 "Bounceless" switch.

Figure 4–8 shows another circuit that can be used to "debounce" pushbuttons; it is known as the *Schmitt trigger.* It is similar to a monostable multivibrator but operates in a bistable manner; the output changes state when the input changes state. Another important use of a Schmitt trigger is to take an analog waveform (such as a sine wave) and produce a stream of digital pulses. Although implemented here using NAND gates, there is a TTL IC, the 7414, which comes with six Schmitt trigger circuits in one package.

The principles of oscillation work with logic gates just as well as they do with op amps or other linear ICs. The output of an AND gate (let's call it gate 2) can be used as one of the inputs of another NAND gate (gate 1); in turn, the output of gate 1 serves as one of the inputs for gate 2. If the output of gate 1 switches states, the output of gate 2 can be made to switch states and the output of gate 2 will then cause the output of gate 1 to switch states. In this manner, an oscillation is set up. The output of digital circuits,

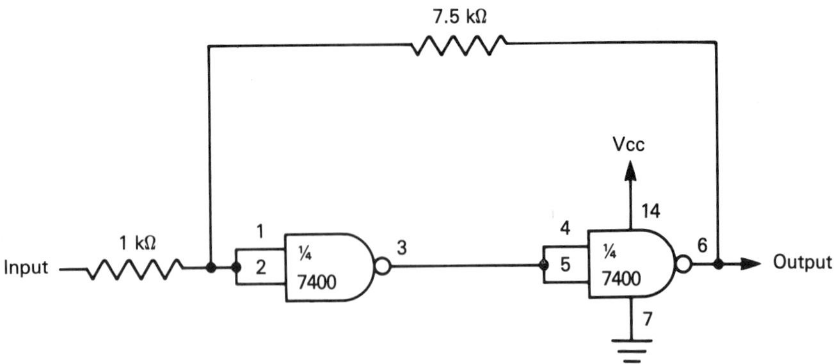

Figure 4–8 Schmitt trigger using the 7400.

however, is a stream of pulses rather than a continuous waveform. (However, if the output of such an oscillator is used to drive a speaker, it will be heard as an analog waveform if the oscillation frequency is high enough.) Figure 4–9 shows a type of oscillator, known as a *clock generator,* which

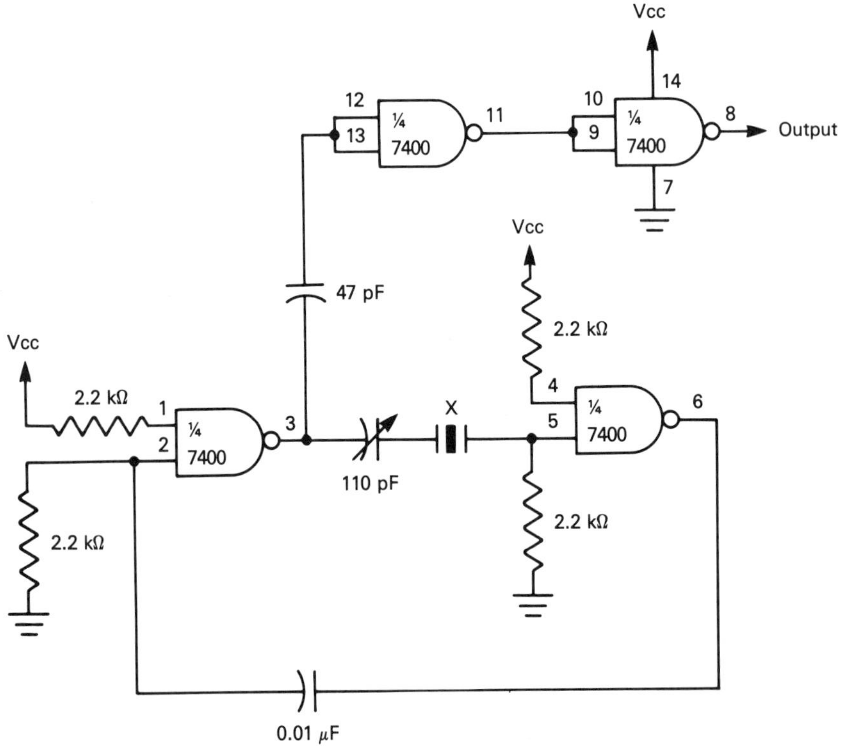

Figure 4–9 Clock generator built from NAND gates.

makes use of "digital feedback." The output frequency depends on the resonant frequency of the crystal X and can range from 100 kHz to 3 MHz. X is "cut" to the desired operating frequency, and the output frequency is adjusted to the correct value by the 110-pF variable capacitor. Placement of components and short leads are critical in this circuit, especially at higher frequencies. If a permanent version of this circuit is assembled, a crystal socket can be used to allow crystal substitution for different operating frequencies. A circuit such as this can provide clock signals for sequential circuits.

The applications given in this section are only a fraction of those possible using the NAND gates of the 7400. Experiment with your own designs by combining the circuits that we have discussed; as long as you observe the operating rules for all TTL devices, the possibility of damaging a device is minimal.

Various other NAND gate devices are available. The 7410 contains three three-input NAND gates in a single package, while the 7420 offers two four-input NAND gates. If those are not enough inputs for you, consider the 7430—that device contains one *eight-input* NAND gate in its package.

THE 7404 HEX INVERTER

A "hex inverter" may sound like a device to reverse the effects of an evil witch's spell, but it is actually a name given to ICs containing six NOT gates, also known as *inverter* circuits. As you will recall from Chapter 1, a NOT gate or inverter takes a single input and reverses it; a high input becomes a low output, while a low input becomes a high output. Figure 4–10 shows the pin connections and internal diagram for the 7404 hex inverter IC.

At first glance, it might seem that a circuit function as simple as an inverter and having only one input would be of little use in the real world. This is not the case, however. The NOT gate greatly simplifies design work when used with other types of logic gates and can also be used independently to create circuits. One useful example is the *input expander* shown in Figure 4–11. This circuit provides two output signals that are opposite in state (high or low) to an input signal. If it is necessary to have the input and output at the same logic level, each inverter can be followed by another inverter. This will give both outputs the same logic level as the input and, in fact, creates a *buffer* circuit with two outputs.

The circuit in Figure 4–11 and its buffer variation are useful when you need to have a single output gate (such as a NAND gate on a 7400) provide two or more logic input signals for other devices. More than two outputs can be obtained by "chaining" additional expander or buffer circuits to either or both of the inputs in Figure 4–11.

Schmitt trigger circuits can easily be implemented using inverters, as shown in Figure 4–12. Note the similarity between this circuit and the Schmitt

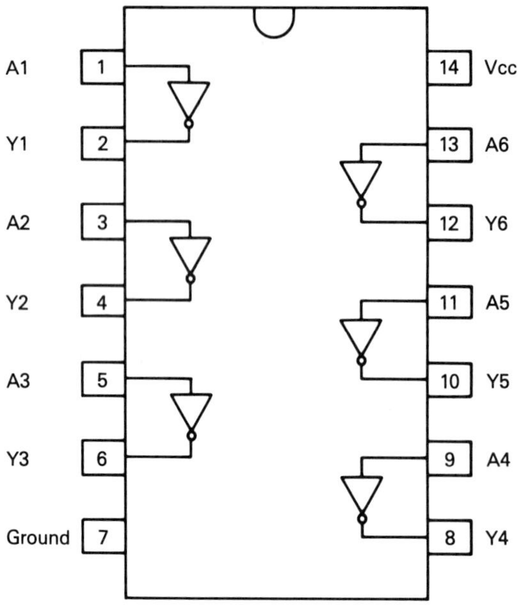

Figure 4–10 Pin connections and internal diagram of the 7404 hex inverter.

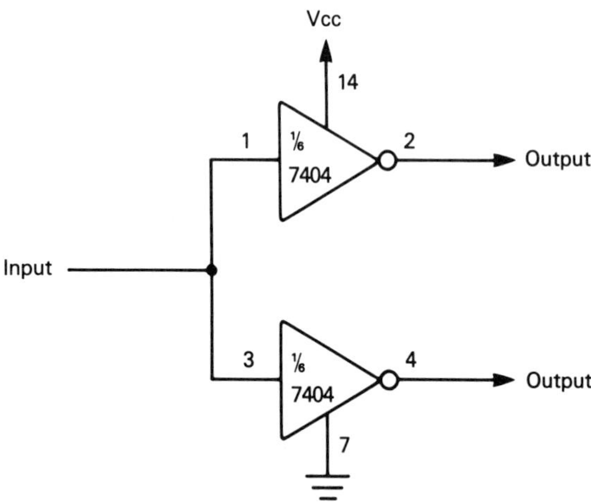

Figure 4–11 Input signal "expander."

trigger implemented with NAND gates shown in Figure 4–8. However, fewer connections are required than with NAND gates since the inverter is already a one-input gate.

Other circuits previously implemented using NAND gates can also be implemented with fewer connections using NOT gates. Figure 4–13 shows

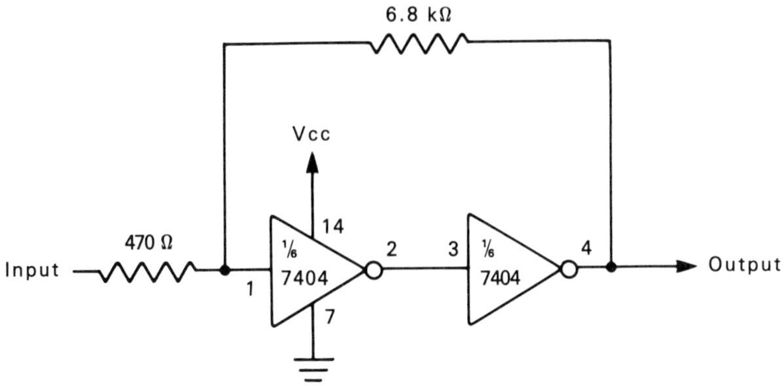

Figure 4–12 Schmitt trigger built from inverters.

a selectable output switch similar in function to the circuit in Figure 4–6; note, however, that the circuit in Figure 4–13 uses only the 7404 IC and a switch. The output can be set to low or high by moving the switch to the position indicated on the schematic.

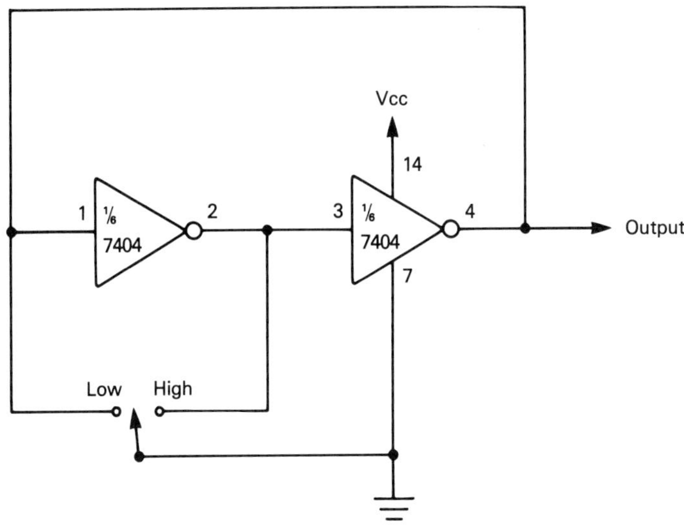

Figure 4–13 Selectable output using inverters.

Figure 4–14 shows two inverters used in a "steady-state" pushbutton circuit. When the button switch is pushed, the output changes state and remains in that new state until the button is pushed in the opposite direction, at which time the output changes back to its original state. The pushbutton switch shown may be replaced by a relay or other switching device, as desired.

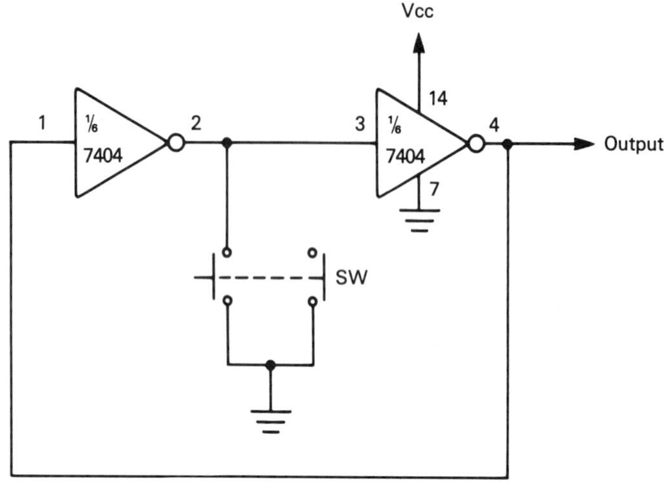

Figure 4–14 "Steady-state" pushbutton.

Using an inverter with NAND gates simplifies the implementation of a type of flip-flop circuit known as the *"D" flip-flop.* Figure 4–15 shows a circuit for one using an inverter and NAND gates. The "D" stands for *delay* or *data.* Both terms are appropriate descriptions, since the flip-flop does not switch states until an enable (or clock) signal is received at input E and there is a single "data" input, D, instead of the reset (R) and set (S) inputs found on those circuits in Figures 4–4 and 4–5. The signal present at input D, when an enable or clock signal is received at input E, determines the output signals at Q and $\overline{Q}$.

When the signal at E is low, the circuit ignores the signal present at input D. When the leading edge of a positive pulse is received at input E, the signal present at that moment at input D controls the status of Q and $\overline{Q}$. Q always has the same logic state as D, while $\overline{Q}$ has the opposite state of Q. If there are no further signals at E, the circuit will "latch" at its current output states and remain that way regardless of changes in the signal at D. This makes the D flip-flop, like the enabled R-S variety, especially suitable as a memory circuit.

Oscillation is easy to achieve with inverters. The circuit in Figure 4–16 has the output of each inverter charging an electrolytic capacitor. As each capacitor charges and then discharges when the inverter's output goes to low, it provides an input signal for the input for the other inverter; this produces an oscillation sufficient to drive an 8-Ω speaker. The circuit in Figure 4–16 produces an output of 4 kHz; the frequency can be altered by changing the values of the 0.1-μF capacitors (both must have the same value to sustain oscillation).

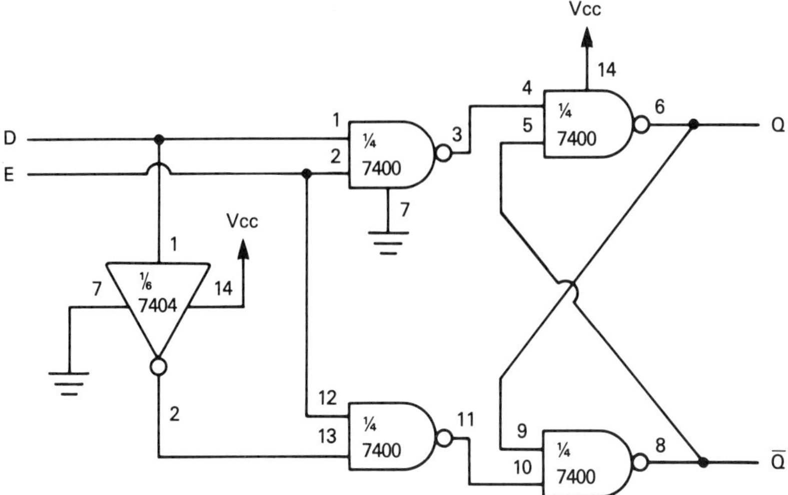

Figure 4–15 D flip-flop built from logic gates.

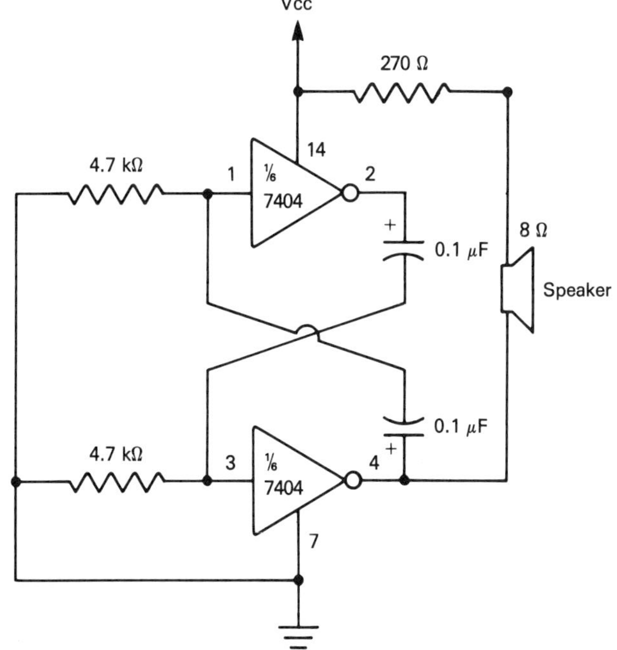

Figure 4–16 4-kHz tone generator using inverters.

THE 7408 QUAD-AND-GATE DEVICE

Figure 4–17 shows the pin connections and internal diagram for the 7408. This device contains four independent two-input AND gates in a single IC package. The function of each AND gate is the same as described in Chapter 1.

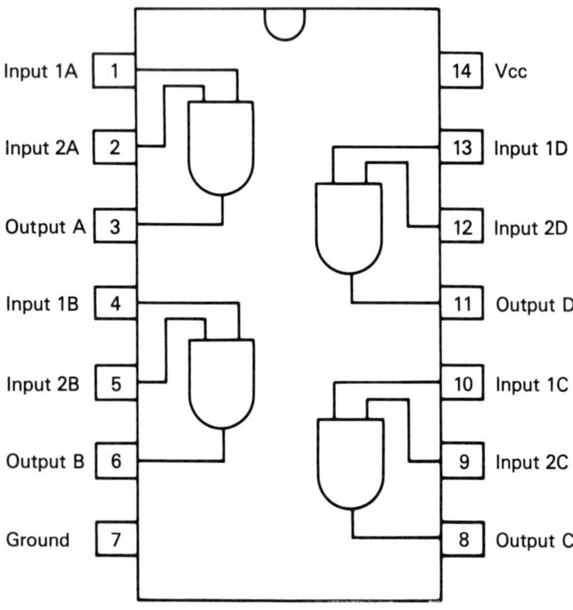

Figure 4–17 Pin connections and internal diagram for 7408 quad-AND-gate device.

Figure 4–18 shows some of the different circuits that can be implemented using AND gates. One of the main uses for AND gates is to add additional inputs for normally two-input circuits such as AND and NAND gates, as shown in Figure 4–18. AND gates make simplified buffers merely by feeding the same signal to both inputs. When its inputs are preceded by inverters, an AND gate will function as a NOR gate. Finally, a switch between one input and Vcc can produce an *enabled AND gate.* When the switch is set to ground, the output of the AND gate will always be low, regardless of the other input's status. If the switch is set to Vcc, a high signal at the other input will produce a high output.

If additional inputs are needed for an AND-gate logic function, the 7411 offers three three-input AND gates and the 7421 gives two four-input AND gates.

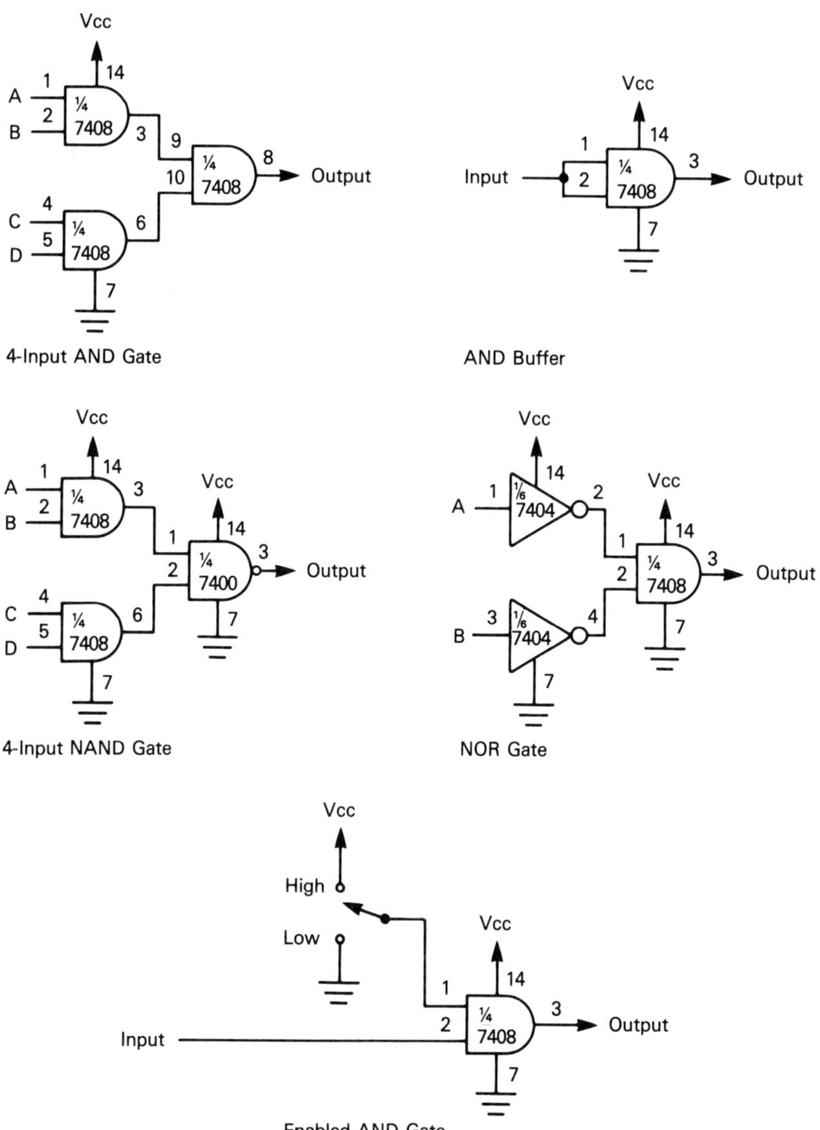

Figure 4–18 Applications of the 7408 AND-gate device.

OTHER TTL LOGIC GATES

Some of the gates implemented earlier in this chapter using AND, NAND, and NOT gates are available in IC form. For example, the 7432 device contains four two-input OR gates in a single package; its pin connection and internal diagram are shown in Figure 4–19. Figure 4–20 gives the pin connections and internal diagram for the 7402 device, which contains four two-input NOR gates in a single IC package. (If additional inputs are needed, the 7427 device contains three three-input NOR gates.) Finally, Figure 4–21 shows the pin connections and internal diagram for the 7486, which contains four two-input XOR gates in its package. All the gates contained on these devices obey the same rules of logic as discussed earlier, and good practice is to use these instead of implementing the gate function from NAND, AND, or NOT gates.

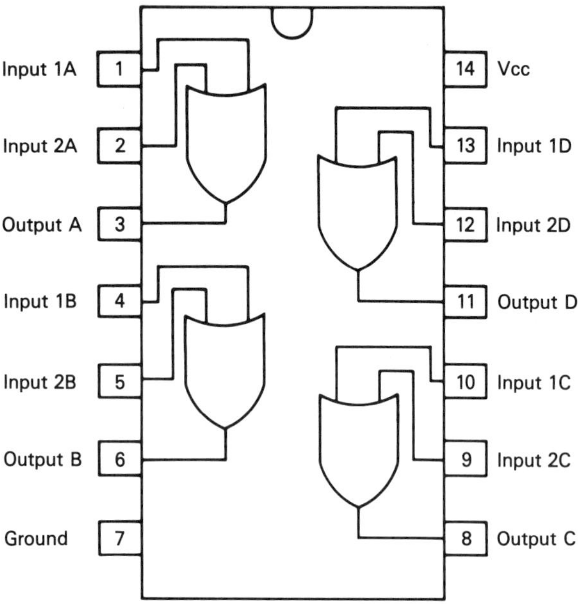

Figure 4–19 Pin connections and internal diagram for the 7432 quad-OR-gate device.

An interesting device is the 74132, whose pin connections and logic diagram is shown in Figure 4–22. It contains four independent Schmitt triggers. A Schmitt trigger implemented with two NAND gates was shown in Figure 4–8. The Schmitt triggers contained by the 74132 are different, since they have *two* inputs rather than one.

You will note that the symbol of a Schmitt trigger is similar to that of a NAND gate. That is because the Schmitt triggers contained on the 74132 are actually NAND gates that have been optimized for use with slowly changing signal levels, such as the analog waveforms that typically make up the input to a Schmitt trigger. The action of these Schmitt triggers is such that the output will be low if *both* inputs are high. If either or both of the inputs

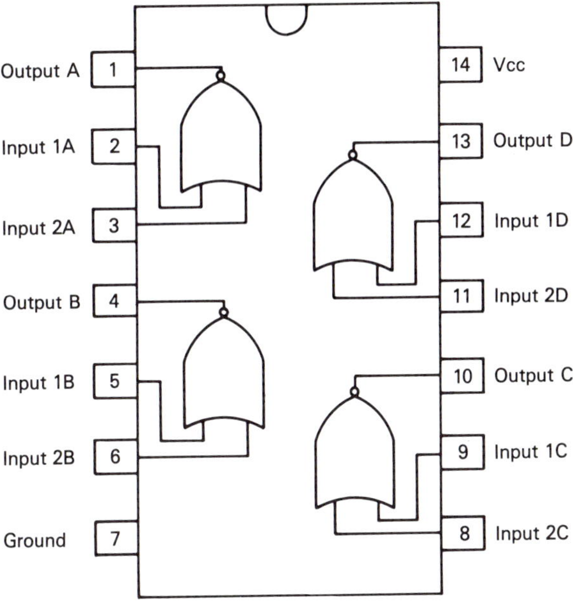

Figure 4–20 Pin connections and internal diagram for the 7402 quad-NOR-gate device.

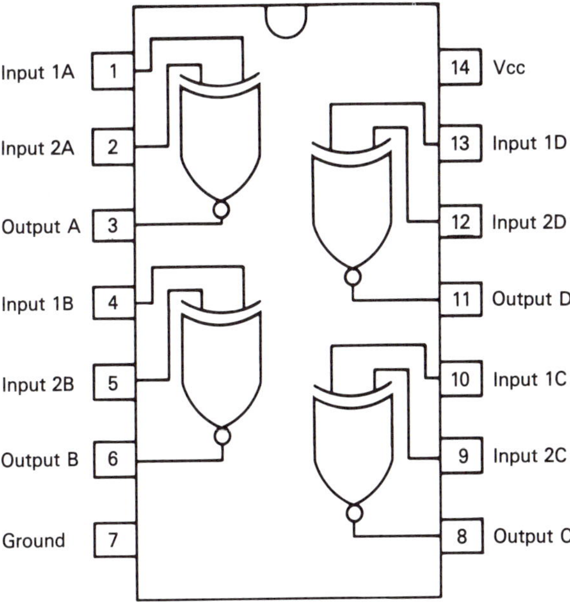

Figure 4–21 Pin connections and internal diagram for the 7486 quad-XOR-gate device.

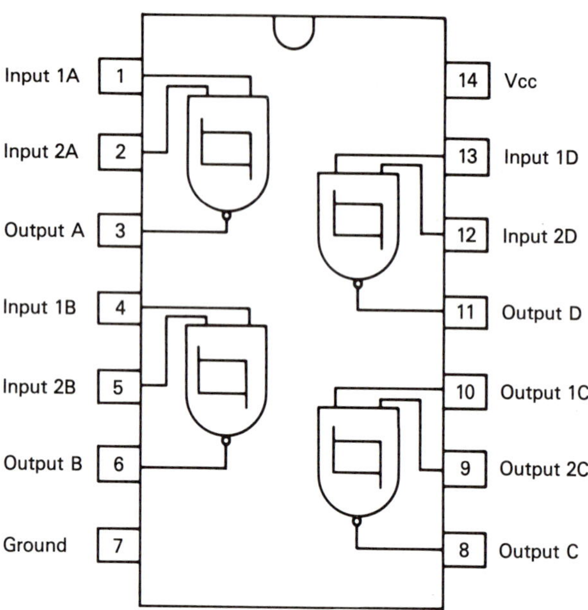

Figure 4–22 Pin connections and internal diagram for the 74132 quad-Schmitt-trigger device.

are low, the output will be high. If only one input is used, the other must be held to high. (Most Schmitt trigger applications involve the CMOS equivalent of the 74132; these applications are discussed in Chapter 5.)

As mentioned earlier, the circuits presented so far are only a fraction of those possible to implement using logic gate devices. As long as the standard operating precautions for TTL are observed, you can experiment with different combinations of logic gates as much as you desire. The only penalty for an improper design will be a nonfunctioning circuit.

THE 74121 INTEGRATED MONOSTABLE MULTIVIBRATOR

Like other circuits, monostable multivibrators can be integrated into a single package. The simplest version is the 74121 monostable multivibrator. Figure 4–23 shows the pin connections and connections of the timing resistor and capacitor for this device.

There are several interesting points about the 74121. For example, note that there are *two* outputs, labeled Q and $\overline{Q}$ (just as with the R-S flip-flop circuits in Figures 4–4 and 4–5). These outputs are, like those of the R-S flip-flop, complementary and must be the opposite state to each other. The 74121 is also a *nonretriggerable* multivibrator. When an input (or trigger) signal is received at one of the inputs, the output changes and remains at that state

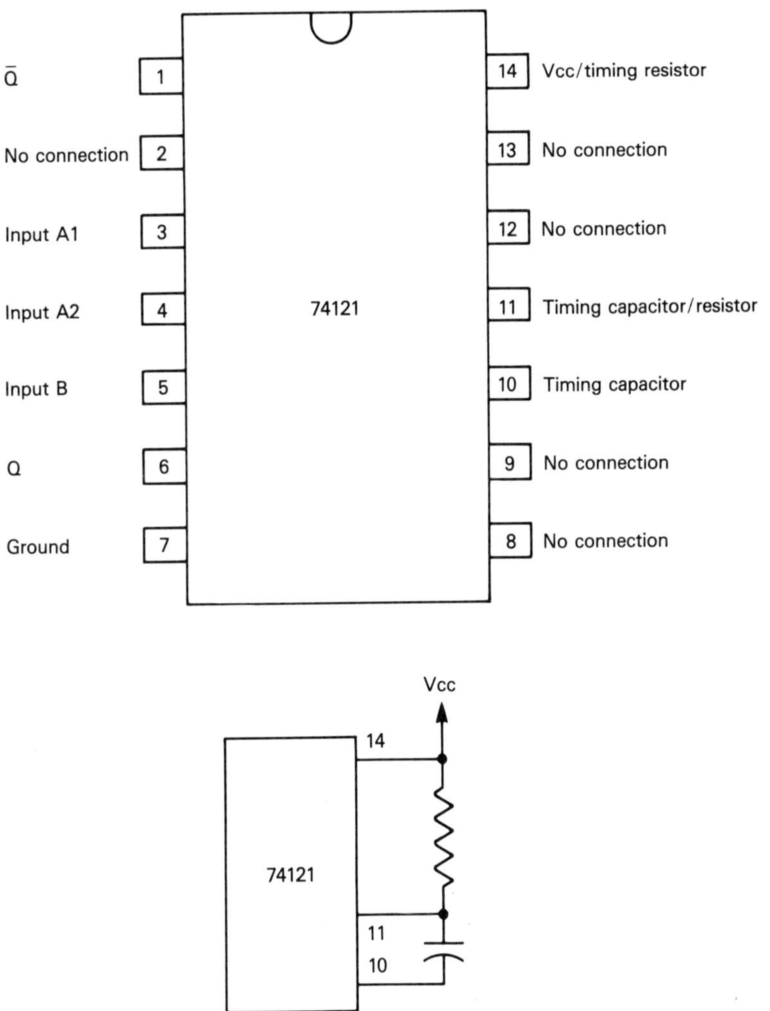

Figure 4–23 Pin connections and timing diagram for the 74121 monostable multivibrator.

for a period of time determined by the timing resistor and capacitor, *regardless* of any changes in the input/trigger during that period. Only after the end of the "output period" will changes in the input affect the output. The trigger signal is a change in the input signal level and is not related to the frequency or repetition of the input signal. Finally, the 74121 can be *negative triggered*. In this situation, the input is normally at a high logic level; when it drops to low, the circuit is activated.

Triggering of the circuit can be accomplished by several methods. If inputs A1 and A2 are connected to ground, the trigger signal at input B involves going from a low to a high logic level. If inputs A1 and B are held to high, dropping a signal at input A2 from high to low triggers the circuit. In a similar manner, if inputs A2 and B are held to high, dropping input A1 from high to low serves as a trigger.

Figure 4–24 shows the 74121 configured as a negative-triggered monostable multivibrator. The value of timing capacitor C can range from 1 μF to 10 pF, while timing resistor R can be in the range 1.6 to 39 kΩ. The exact values of R and C to produce an output of desired duration are difficult to compute and are best determined experimentally. The output duration increases with the values of the timing capacitor and resistor.

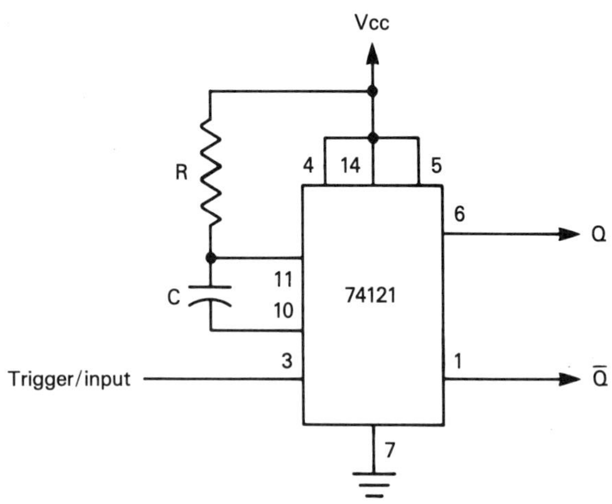

Figure 4–24 The 74121 configured as a negatively triggered multivibrator.

The principal advantage of using the 74121 is the availability of complementary output if needed. It is best suited for very short output times; for longer output times, circuits based on the 555 timer discussed in Chapter 3 are preferable. Circuits using the 74121 and similar TTL multivibrator ICs are prone to improper operation due to false triggering; proper circuit layout, use of bypass capacitors, and "clean" trigger inputs are essential.

Two other popular TTL multivibrators are the 74122 and 74123. The 74122 contains a single multivibrator which may be triggered at any time (termed *retriggerable*), unlike that in the 74121, which needs recovery time after it is triggered before it can be triggered again. The 74123 contains two retriggerable multivibrators.

FUNDAMENTALS OF TTL FLIP-FLOP DEVICES

As mentioned before, the flip-flop circuits implemented earlier using logic gates are also available as integrated devices. Flip-flop circuits also have their own distinctive schematic symbols; these are shown in Figure 4–25. You will see these symbols in circuits using integrated flip-flop devices and in the internal diagrams of such devices.

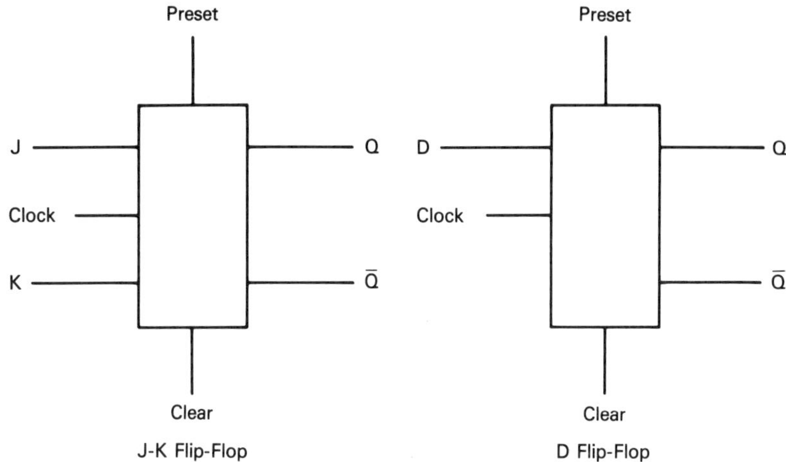

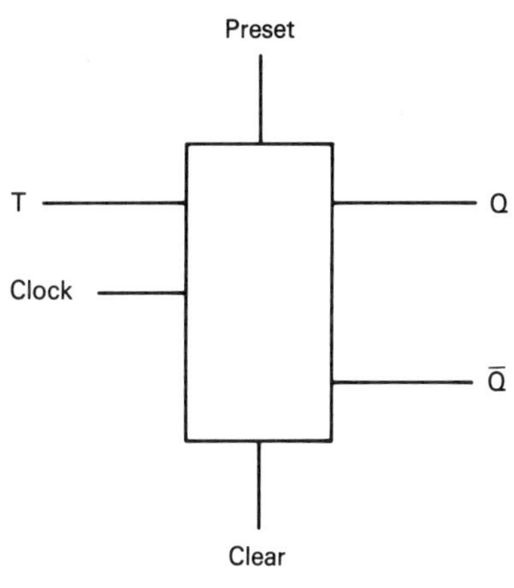

Figure 4–25 Schematic symbols for various flip-flops.

The *J-K flip-flop* is similar to the R-S flip-flop circuit examined earlier; the J input corresponds to the S input and the K input to the R input. What do the "J" and "K" stand for? The origin and meanings of them are actually unknown, but they do help differentiate this circuit from the older R-S flip-flop. Actually, there are several important differences between the two. In function, the J-K flip-flop more closely resembles the enabled R-S flip-flop shown in Figure 4–5 than the "classic" R-S circuit in Figure 4–4. The *clock* input is similar to the enable input in Figure 4–5. Moreover, most (but not all) J-K flip-flops include two new inputs known as *preset* and *clear*. These are included because the outputs, Q and $\overline{Q}$, assume states at random when power is first applied to the circuit. The clear input signal is equivalent to a reset (R) input while the present input signal is the same as a set (S) input. The preset and clear inputs are used at times other than initially applying power; for example, in arithmetic circuits they may be used to enter initial values or clear the count to zero.

The preset and clear inputs are special in that they are *asynchronous*. This means that they affect the circuit regardless of the clock input signal. (The effects of the J and K inputs depend on the clock signal, and thus are *synchronous* inputs.) Thus attention must be paid to keeping stray or transient signals off the clear and preset inputs.

For the J-K flip-flop shown in Figure 4–25, we present below a logic table showing the outputs depending on the preset (*pre*), clear (*clr*), clock (*clk*), J, and K signals. In this table the symbol "X" is used to indicate that the state of the signal—whether high or low—is irrelevant. In addition, the symbol "E" is used to indicate that a clock signal ("enabling" the circuit) is being received at the clock input.

Pre	Cir	Clk	J	K	Q	$\overline{Q}$
L	H	X	X	X	H	L
H	L	X	X	X	L	H
H	H	E	H	L	H	L
H	H	E	L	H	L	H
H	H	E	H	H	Outputs toggle in response to incoming clock pulses	

From the table it is apparent that the clear and preset inputs must be held to a high logic level if normal operation of the circuit is desired.

Figure 4–25 also shows the schematic symbol for a D flip-flop. This is very similar to the J-K flip-flop, except that there is only input, D, rather than two (J and K). The clear and preset inputs function as they do for the J-K circuit, and the table of its operation is as follows:

Pre	Clr	Clk	D	Q	$\overline{Q}$
L	H	X	X	H	L
H	L	X	X	L	H
H	H	E	H	H	L
H	H	E	L	L	H

The final schematic symbol in Figure 4–25 is for a *T flip-flop* (sometimes known as *R-S-T flip-flop*). The "T" stands for *toggle,* and as the name implies, the outputs change states in response to the T input. In practice, a T flip-flop is formed by connecting the inputs of a J-K flip-flop. Currently, no T flip-flop ICs are available (since they are so easy to implement from J-K flip-flops), but these circuits sometimes show up in the internal diagrams for other ICs, such as counters. The function of a T flip-flop is simple; each time the T input is high and the circuit is enabled, the outputs toggle.

You will often see a J-K or D flip-flop schematic symbol similar to the one in Figure 4–26. At first glance it appears similar to the J-K flip-flop symbol in Figure 4–25. However, notice that the clock, preset, and clear inputs terminate in small circles when they meet the body of the device. This indicates that the device uses *negative* or *inverted triggering.* The signal at the clock input is normally high, and a clock signal is provided when the input drops from high to low (a negative pulse). The inverted clear and preset inputs operate in the opposite fashion from a "normal" flip-flop; a low preset signal sets the circuit, and a low clear signal clears the flip-flop.

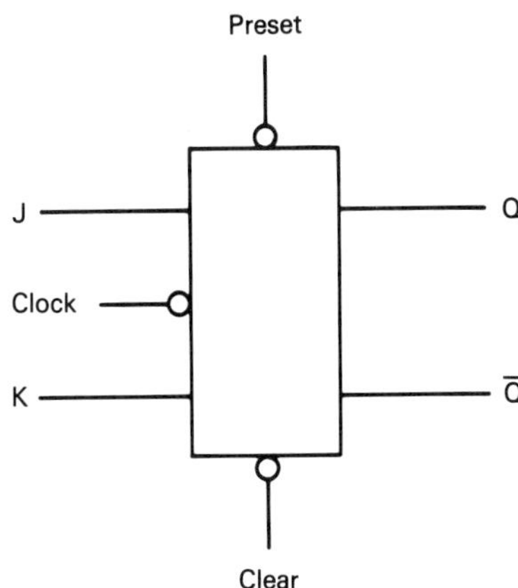

Figure 4–26 Schematic symbol illustrating inverted trigger, clear, and preset inputs on a J-K flip-flop.

It is possible for only one input, such as the clear or clock input, to have inverted or negative triggering. Some flip-flop ICs are designed in this fashion, and it can be implemented easily by preceding an input with an inverter.

So far, we have used the term "clock" and its synonyms without precisely defining it. It so happens that there are actually two different types of clock signals. *Level clocking* involves the state (high or low) of the signal itself acting as the clock. It must be held at the triggering level long enough for the circuit to "latch." By contrast, *edge clocking* is where the *change* in the clock signal level (from low to high or high to low) provides the clock action. The duration of the clock signal does not matter—only the beginning or end of the pulse. The beginning edge of a clock pulse is known as the *leading* edge and the ending edge is the *trailing* edge.

These may seem like minor distinctions, but they have some important real-world implications. If a device uses level clocking, the input signals applied when a clock pulse is received must be removed before another set can be applied. The inputs cannot change except after a clock pulse, and then only once. With edge triggering, the inputs may change at any time and the proper output will result. J-K and T flip-flops are normally level clocked while D flip-flops are edge clocked. Flip-flop ICs are identified as to the clocking method used. Failure to note the proper clocking system and the differences in how input signals are handled are among the most frequent errors in designing and building circuits using flip-flop ICs.

USING TTL FLIP-FLOP ICS

Figure 4–27 shows the pin connections for the 7473 device. This IC contains two independent J-K flip-flops in a single package. Note that both flip-flops have a clear input, but *do not have a preset input.* This means that you can clear the flip-flops at any time, but a preset value cannot be entered into a circuit where they are the main computational or storage elements. (Also notice that the Vcc and ground pins are located at different pins than those used on devices discussed previously in this chapter.) Since the flip-flops contained by the 7473 are of the J-K variety, they are level triggered.

The following logic table describes the operation of each flip-flop in the 7473:

Clr	Clk	J	K	Q	$\overline{Q}$
L	X	X	X	L	H
H	E	H	L	H	L
H	E	L	H	L	H
H	E	H	H	Outputs toggle in response to incoming clock pulses	

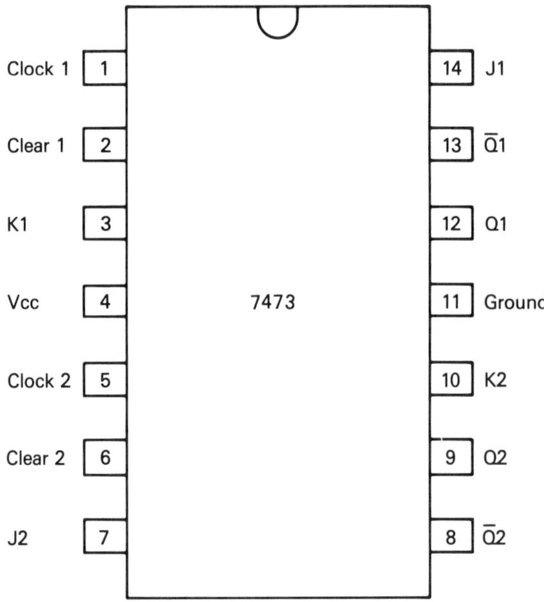

Figure 4–27 Pin connections for 7473 dual J-K flip-flop with clear input.

One useful application of the 7473 is to construct *binary counter* or *divider* circuits. Two examples of such circuits are shown in Figure 4–28. The divide-by-2/counter circuit shown makes use of one of the 7473's J-K flip-flops. Notice that only one output, Q, is used and that the J and K inputs are both "tied" to Vcc and thus are held at a high logic level. (The clear input must also be held to high, as indicated by "H" in Figure 4–28.) As the logic table above indicates, under such conditions the flip-flop will toggle in response to each clock pulse.

Each clock signal will cause Q and $\overline{Q}$ to switch states. Thus the output at Q will be high on every other clock pulse; during the other pulse, $\overline{Q}$ will be high. However, $\overline{Q}$ is not used in this circuit, meaning that a high output will be available (at Q) on only *half* the input pulses. If the clock frequency is 100 Hz, a high output will be available from this circuit at only half that rate, or 50 Hz. In effect, the clock frequency has been "divided" in half. (What really happens, of course, is that the other half of the high-level outputs available at $\overline{Q}$ have been discarded.)

It is also possible to view this circuit as an *automatically recycling counter*. It is admittedly a limited counter—it can only count 0 and 1—but it does indeed count the incoming clock pulses. Suppose that the output has been set to 0 by a clear input signal. The first clock pulse will send the output high (the first clock pulse is counted) while the second sends the output low (the second pulse is counted). The third clock pulse again sends the out-

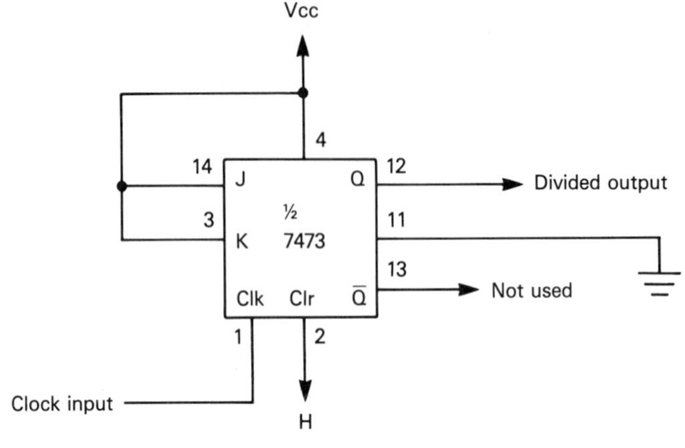

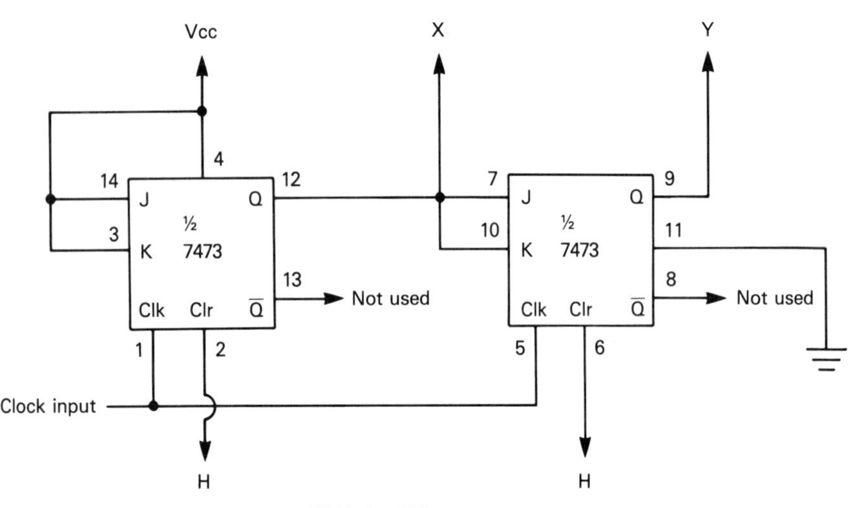

Figure 4–28 Division/counter circuits using the 7473.

put high, and the counting cycle is then repeated. Thus a counter can be defined as a circuit that produces one output pulse for a given number of input signals.

Such a circuit is not very useful. However, it is possible to chain several such circuits together to count higher. The divide-by-4 and counter circuit in Figure 4–28 is an example. Both "halves" of the 7473 are used, with both sharing a common clock signal. When used as a divider, the first flip-flop produces an output that is half the clock frequency. This output in turn serves as the J and K inputs for the other flip-flop. The result is that X and Y can have four different outputs. Both start at low, and then X goes high while Y remains low. Then they toggle, as Y goes high while X goes low. Finally, both go to high and then "recycle" back to both being low. The clock signal is thereby divided into four possible states.

This also means that the circuit can count from 0 to 3 (or, in binary terms, from 0 to 11). The counter starts at 0 (00), and the next clock pulse makes the count 1 (01). The next pulse makes the count 2 (10), with the following pulse incrementing it to 3 (11). That is the limit of the circuit, and the next pulse sends it back to 0.

The operation of the circuits in Figure 4–28 can easily be visualized by adding LEDs to the outputs. Figure 4–29 shows how LEDs can be added to light either on a low or a high output. LEDs can also be added to the clock signal inputs using the high output diagram to give "before and after" comparisons of input and output. (The methods illustrated in Figure 4–29 will also work with the outputs of all TTL devices discussed in this chapter.)

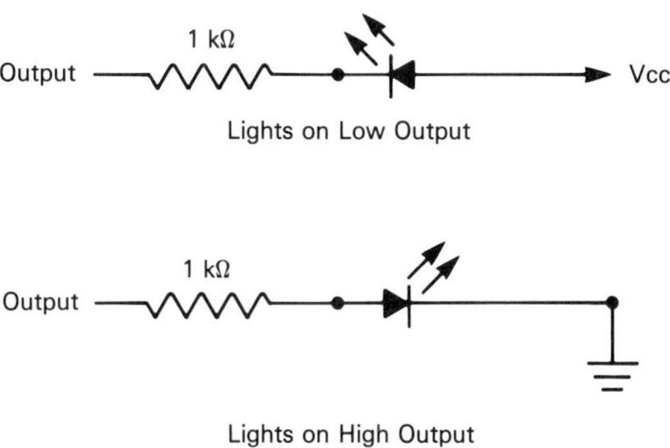

Figure 4–29 Methods of using LEDs with TTL outputs.

Figure 4–30 shows the pin connections for the 7476 device. This IC contains two J-K flip-flops, with each having a clear and preset input. The logic table for each flip-flop is the same as that previously given for the J-K flip-flop in Figure 4–25. The important point to remember is that both the preset and clear inputs must be high for the circuit to operate, and that holding all inputs—preset, clear, J, and K—to high will cause the outputs to toggle in response to incoming clock pulses.

The chief advantage of the 7476 over the 7473 is the addition of preset inputs. The 7476 can be used in the same circuits as the 7473 (such as those shown in Figure 4–28) as long as both the clear and preset inputs are high. The preset inputs, however, make it suitable for use in a circuit known as a *shift register.* Figure 4–31 shows a shift register using four J-K flip-flops

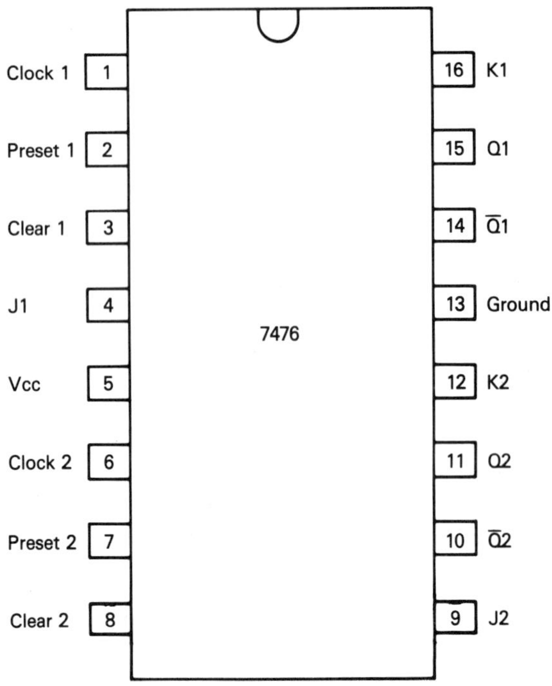

Figure 4–30 Pin connections for 7476 dual J-K level-triggered flip-flop with preset and clear inputs.

from two 7476 devices. (To save space, the clear and preset inputs have not been labeled; they are numbered, though.)

As you can see from Figure 4–31, a shift register is a series of flip-flops connected so that the outputs of one flip-flop serve as the inputs for the next flip-flop in series. This is called *cascading* flip-flops. All flip-flops receive the same clock signal simultaneously. The shift register gets its name from the fact that data entering from the input on the left are ''shifted'' to the right on each clock pulse. Shift registers can be used to convert data from serial form (one bit after another on a single line) to parallel form (each bit has its own separate line); they can also convert parallel data to serial form. They can also be used as memories. The circuit in Figure 4–31 is a 4-bit shift register, with serial input converted to parallel form present at outputs A, B, C, and D. The input is directly applied to the J input of the leftmost flip-flop and simultaneously through an inverter (from a 7404 hex inverter IC) to input K. This arrangement ''splits'' the single-line serial input into two complementary inputs for the flip-flop chain. The outputs making up the 4-bit parallel output are taken from the Q output of each flip-flop.

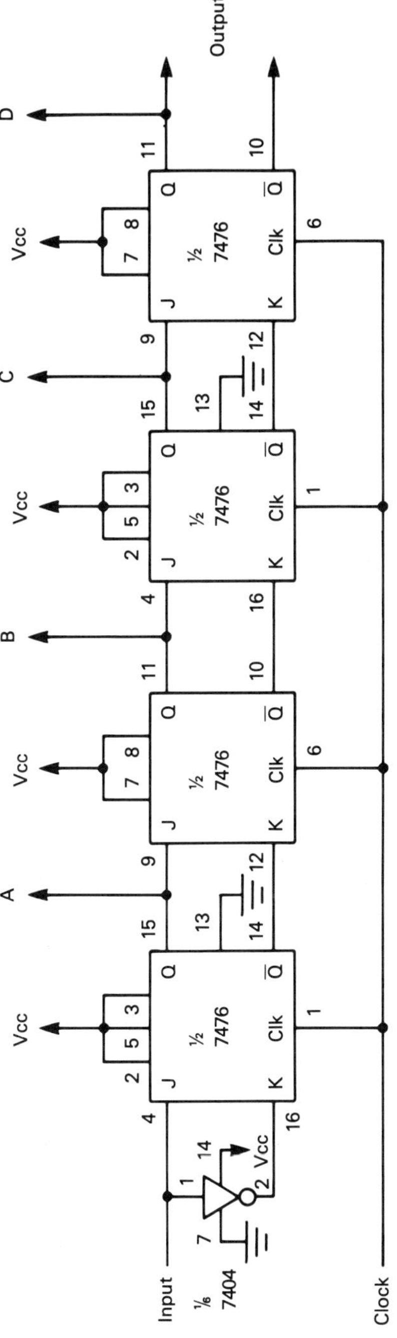

Figure 4–31 Shift register using the 7476.

Shift registers may also be implemented using D flip-flops. Figure 4–32 shows "generic" D flip-flops being used to implement a circuit similar to that in Figure 4–31. (For simplicity's sake, the clear and preset inputs have been deleted, but keep them in mind if you build this circuit.) J-K flip-flops are generally preferred for shift register applications, however, since circuits using J-K flip-flops are generally significantly faster than those using D flip-flops.

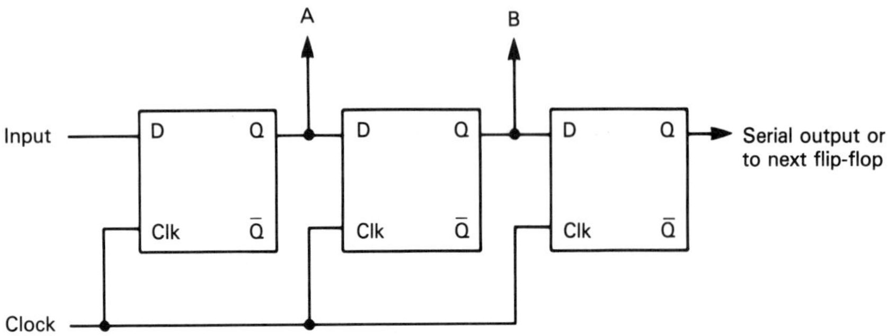

Figure 4–32 Shift register implementation using D flip-flops.

Shift registers are available in IC form and will be examined in a few more pages. These can handle data in 4- or 8-bit "words." Most shift registers shift to the right; data enter on the left side of the circuit and are shifted toward the right on each clock pulse. Some IC shift registers are *bidirectional,* meaning that data can enter from either the left or right "side" of the circuit and be shifted toward the opposite direction.

The shift registers in Figures 4–31 and 4–32 are examples of *serial input, parallel output* (SIPO) shift registers. There are three other types: *serial input, serial output* (SISO), *parallel input, serial output* (PISO), and *parallel input, parallel output* (PIPO). These are available in IC form.

The most popular D flip-flop IC is the 7474. This device contains two independent D flip-flops with preset and clear inputs in the same package. Each follows the same logic table as given for the flip-flop in Figure 4–25. Like other D flip-flops, these are edge triggered rather than level triggered. Figure 4–33 shows the pin connections for this device.

Data present at the D input can be changed anytime, regardless of the clock signal state (high or low). For normal operation, both the clear and preset inputs must be high. With both preset and clear high, the Q output state is an "echo" of the D input level; a high D input produces a high Q output. Although D flip-flops tend to consume less power than J-K types, they are not as versatile. For example, the togg'e condition (outputs switching states in step with clock pulses) is not available.

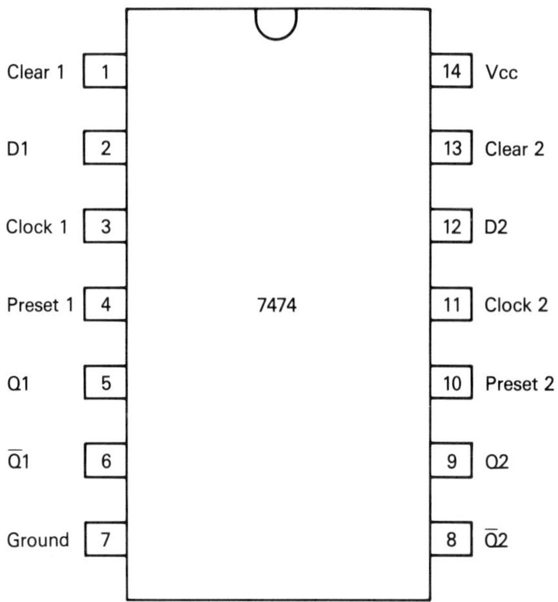

Figure 4–33 Pin connections for 7474 dual D edge-triggered flip-flop with preset and clear inputs.

Figure 4–34 shows two interesting applications of the 7474's D flip-flops. The divide-by-2 circuit produces an output at Q whose frequency is half that of the clock signal. Note that the D input signal is actually the $\overline{Q}$ output "looped back" to D. This causes the output to change states on every positive clock pulse edge. Since the only output available is Q, its frequency is exactly half that of the clock.

The second circuit in Figure 4–34 is a "pseudo" dual-input D flip-flop. Two separate square-wave input signals are applied, with one going to the D input and the other serving as the clock signal. The $\overline{Q}$ output is not used, while the Q output has a signal that is input 2 minus input 1. A circuit such as this can be described as a *digital mixer*. This circuit is also sometimes known as a *down counter,* since the signal at input 2 is reduced by the amount at input 1.

The digital mixer circuit can also be adapted to indicate when the frequencies of two digital signals are or are not equal. The indicating LED connections shown in Figure 4–29 can be added to its Q output. If connected to light when the output is low, the LED will glow when the two inputs are unequal. If connected to light when the output is high, the LED glows when they are equal. Since the outputs are complementary, another LED could be connected to the $\overline{Q}$ output to give both actions and a "lit" indication of the frequencies of the two input signals.

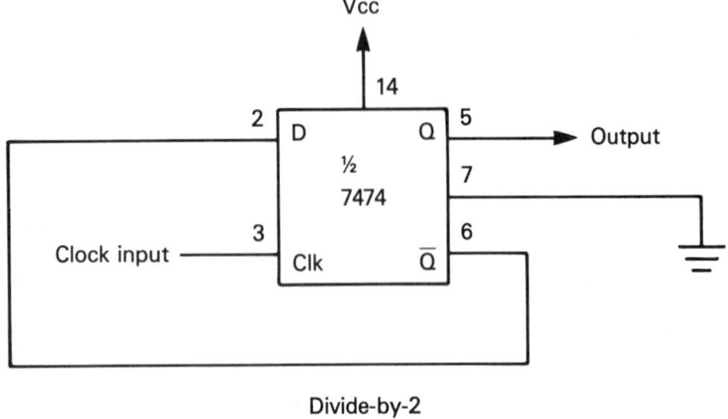

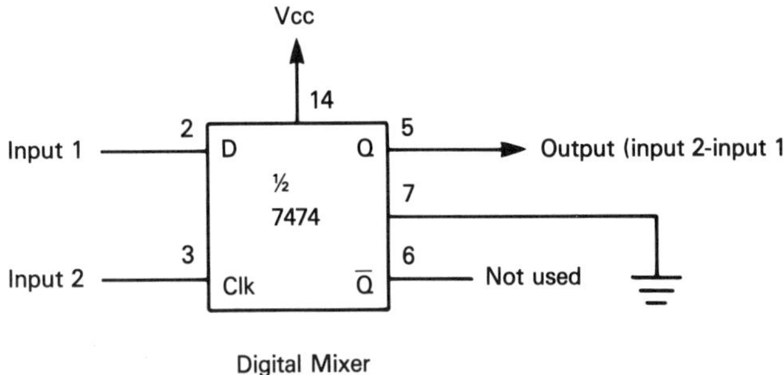

Figure 4–34 Applications of the 7474.

Figure 4–28 showed divider/counter circuits made from J-K flip-flops on the 7473 device. It is also possible to construct similar circuits using D flip-flops, as shown in Figure 4–35. Like the circuits in Figure 4–28, the circuit in Figure 4–35 uses the clock signal as its input, and its output is a frequency that is only a fraction of the clock frequency. The output frequency of this particular circuit is only one-eighth of the clock frequency, so it is a *divide-by-8* circuit. The circuit is also a *down counter.* This means that the count starts at its maximum value (111 in this case) and proceeds *downward* with each clock pulse to 000. A counter like those shown in Figure 4–28 is known as an *up counter.*

Several other TTL flip-flop devices are available. Among them are quad devices with four flip-flops in a single package (such as the 74175) and octal ICs with eight flip-flops (74374, etc.). Almost all flip-flop ICs with four or eight circuits per chip contain D flip-flops.

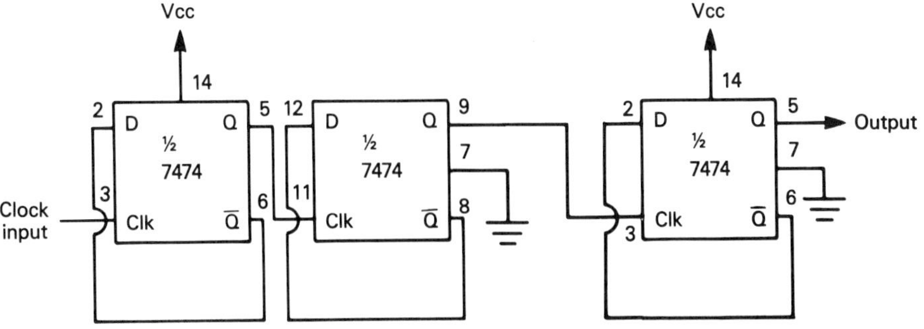

Figure 4–35 Down counter/divider using the 7474.

COUNTER DEVICES

Counters/dividers are available in IC form. These are extremely common in everyday life—for example, digital clocks, watches, and timers are essentially counting circuits based on counter ICs. There are numerous counter ICs currently in use, and in this section we examine two of the more popular devices.

There are some important terms used to describe IC counters. One is its *modulo*. This simply means the range over which it counts before recycling back to zero. A "modulo 10" or *decade* counter has a range from 0 to 9 before recycling; it can also be thought of as dividing the input signal by 10. Some counter ICs have a larger modulo (such as 16). IC counters may also be described as *ripple* or *synchronized.* In a ripple counter, the flip-flops in the counter are individually "clocked" by the output of the immediately preceding flip-flop. The clock signal can be said to "ripple" through the counter, much like a wave caused by tossing a stone into water. In a synchronized counter, all flip-flops are clocked simultaneously by an independent clocking signal. The major practical difference between the two types is that a ripple counter is slower than a synchronous circuit, due to the "ripple through" time.

Counters are often described as being *binary* (hexadecimal) or *decade (BCD).* A decade or BCD (for *binary-coded decimal*) counter is a base-10 counter, which can count from 0 to 9. A binary or hexadecimal counter is a base-16 counter, which counts from 0 to 15 (or 0 to F in the hexadecimal system). A few base-12 counters are also available; these often consist of independent divide-by-6 and divide-by-2 counters in a single package. To count to greater ranges, some counters may be cascaded together. However, not all counters may be cascaded. Counters are also classified by their counting direction; different types can count up, down, or in both directions.

Figure 4–36 shows the pin connections for the popular 7490 decade counter. This device is actually two separate counters, a divide-by-2 and a divide-by-5, in a single package. It is a ripple counter, and counts *only* in an upward director. It is edge triggered, but triggers on each leading *negative* clock edge.

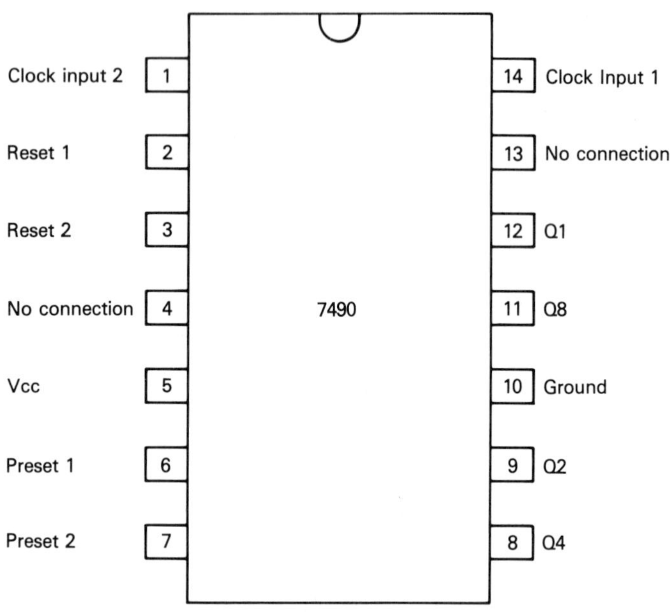

Figure 4–36 Pin connections for the 7490 decade (divide-by-10) counter device.

Figure 4–37 shows the 7490 used as a divide-by-5, divide-by-7, and divide-by-10 counter. Note that *all* reset pins must be connected to ground to use the device as a decade counter. (Also note the unusual locations of the Vcc and ground pins.) The divide-by-5 clock input is clock input 2 (pin 1) and the divide-by-2 clock input is clock input 1 (pin 14). The counter may be cleared by applying a high-level signal to reset input 1 or reset input 2, or both. The similar method can be used to preset the counter using either or both preset input 1 and preset input 2. *The 7490 may not be cascaded.*

A far more versatile counter is the 74192, whose pin connections are shown in Figure 4–38. This is a synchronous up/down decade counter and may be cascaded. Clocking is by the positive leading edge of the clock signal. The direction of the counting is controlled by the down-count input (pin 4) and the up-count input (pin 5). For upward counting, the clock signal is applied to the up-count input while the down-count input is held high. For down-

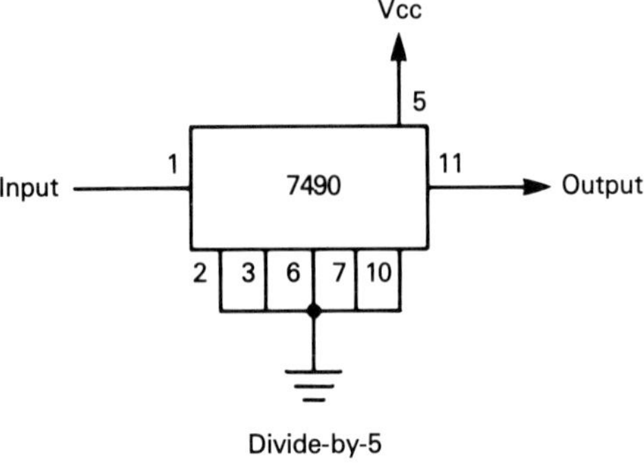

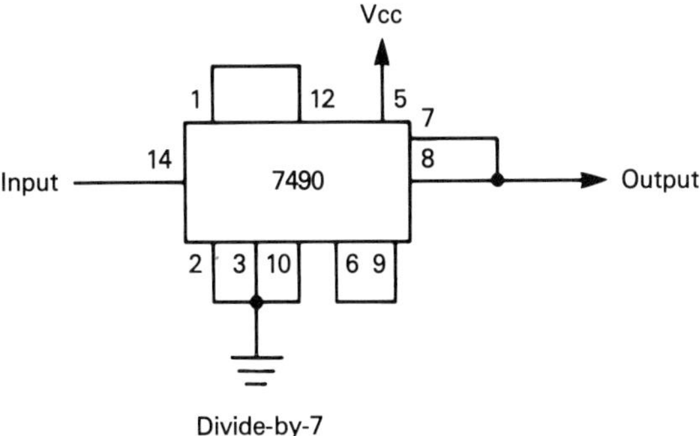

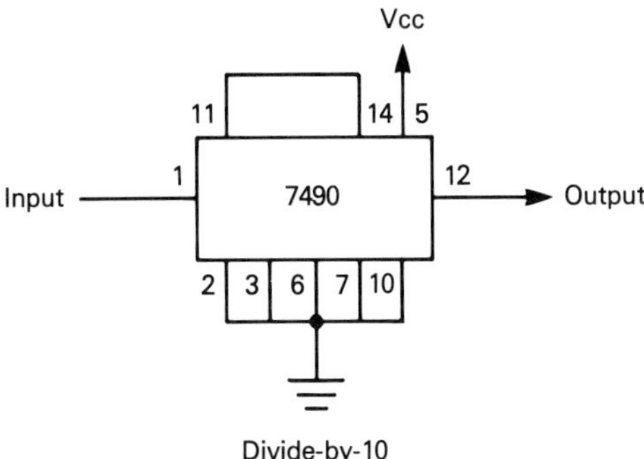

Figure 4–37 Applications of the 7490.

ward counting, the clock signal is applied to the down-count input while the up-count input is held high.

The outputs of the 74192 are labeled Q1, Q2, Q4, and Q8, while the inputs are labeled A, B, C, and D. To load data into the counter, they are "placed" at the inputs and the load input is momentarily dropped to low.

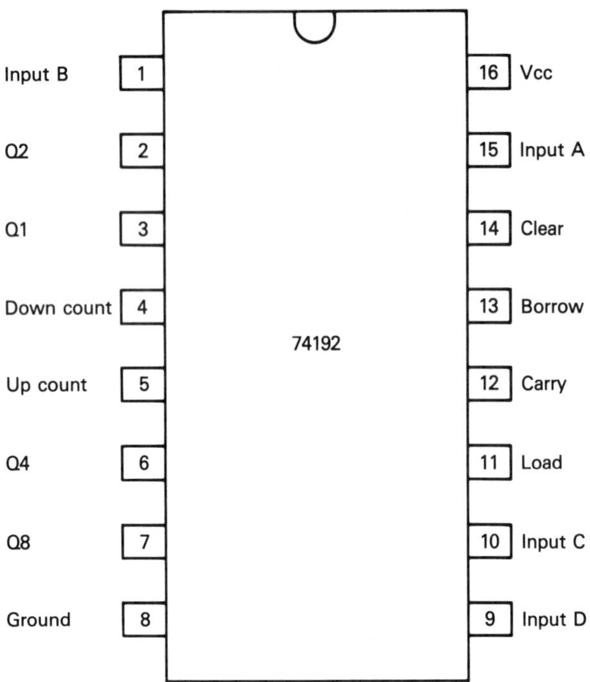

Figure 4–38 Pin connections for the 74192 decade up/down counter.

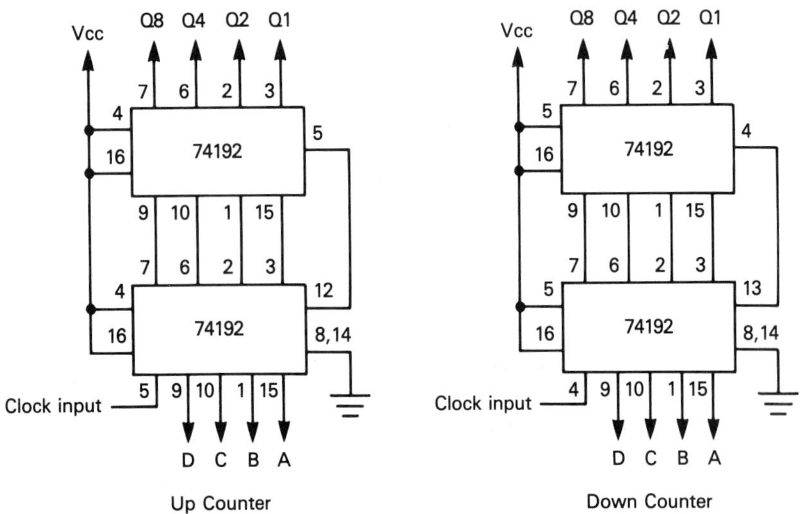

Figure 4–39 Applications of the 74192.

The clear input must be low during normal counting operation; clearing the counter is accomplished by bringing the clear input to high. To cascade more than one counter, the carry output of the first counter is connected to the up count of the second (this produces an up counter) or the borrow output of the first to the down count of the second (this produces a down counter). Figure 4–39 shows two 74192 ICs used to create up and down counters in the range from 0 to 99.

MISCELLANEOUS TTL DEVICES

As mentioned earlier, shift registers are available in integrated form. One of the most versatile is the 74194 4-bit shift register, whose pin connections are shown in Figure 4–40. The 74194 is sometimes referred to as a "universal" shift register because of its versatility. This device is capable of shifting data either to the left or right, and has four inputs and outputs available.

The direction is which data are shifted depends on the signals applied to shift 1 (pin 9) and shift 2 (pin 10). If shift 1 is high while shift 2 is low, the data are shifted right. If shift 1 is low while shift 2 is high, the data are shifted left. Serial data to be shifted right can be input at pin 2 and data

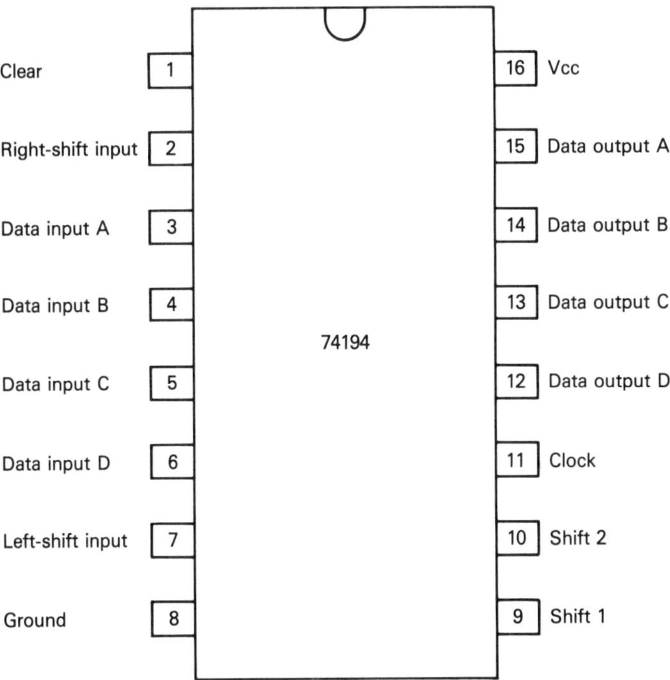

Figure 4–40 Pin connections for the 74194 4-bit shift register.

to be shifted left are input at pin 7; data may also be loaded in parallel at data inputs A through D. Data that are at the parallel inputs when both shift 1 and shift 2 are high are advanced with each clock pulse. The clear pin must be held high during the shifting operation.

Numerous shift registers are available in both 4-bit and 8-bit sizes. Most only allow right-shift operation, have a clear pin, and permit parallel input. Most with parallel input use synchronous loading. The operating principles and applications of these are the same as for shift registers using individual flip-flop devices.

An interesting device is the 7485 4-bit magnitude comparator. This IC takes two 4-bit "words," compares them, and indicates which is larger or if they are equal. The two words are labeled A and B; outputs are provided for A being greater than B (pin 5), B being greater than A (pin 7), and A and B being equal (pin 6). A high output appears at the correct output while the other two go low. The inputs at pins 2, 3, and 4 are used to cascade devices to compare larger words, of 8 bits or greater; otherwise, the A = B input is kept high while the A > B and A < B inputs should be connected to ground. Figure 4–41 shows the pin connections for this device.

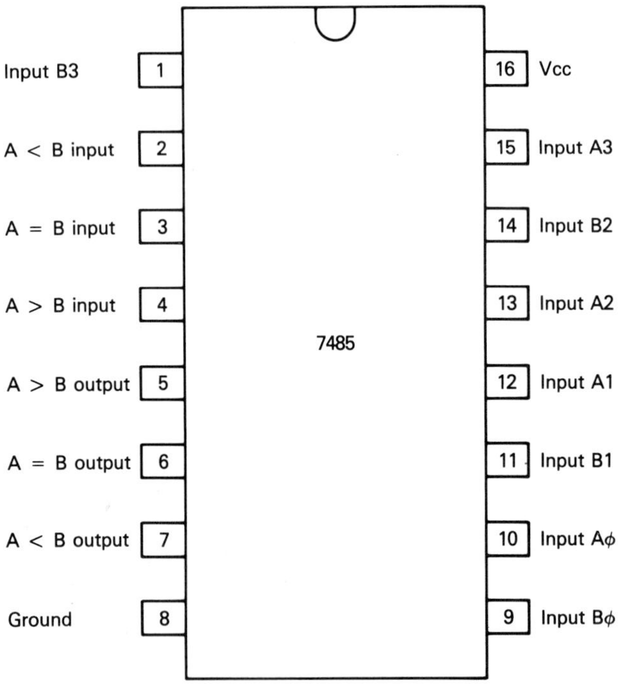

Figure 4–41 Pin connections for the 7485 4-bit magnitude comparator device.

Many other TTL devices have been developed for use in microprocessor and microcomputer systems. These include *data selectors* and *data decoders,* which are used to choose and route data along data buses. There are also *bus transceiver* devices, which receive data from a bus and transmit data to the bus. Although extremely useful, the applications of these devices are outside the scope of this book. The manufacturer's data sheet or equivalent reference will supply the necessary information to use these devices, since their internal circuitry is almost always some combination of the circuits (flip-flops, logic gates, etc.) discussed in this chapter, and the device's function could be implemented using other standard TTL ICs.

chapter 5

CMOS DIGITAL CIRCUITS

CMOS (complementary metal-oxide silicon) digital ICs are now the preferred choice over TTL in most applications. The reasons for this are numerous. Chief among them is the very low power consumption of CMOS compared to TTL, making it ideal for battery-powered applications. The low power consumption means that CMOS is less prone than TTL to generate power supply glitches. Moreover, the transition level between high and low logic states is usually approximately half the supply voltage. This means that CMOS has a higher level of noise immunity than that of TTL.

Another virtue of CMOS is the wide power supply voltage over which it can operate, which is typically +3 to +15 V. (To differentiate between supply voltage requirements of CMOS compared to TTL, the positive supply voltage for CMOS is known as *Vdd* and the ground connection is sometimes labeled *Vss*.) This means that both TTL and CMOS can be powered from a +5-V supply. However, this is not recommended, since the operating speed of CMOS drops with lower supply voltages along with noise immunity. The optimum Vdd for CMOS is usually in the range +9 to +12 V.

CMOS inputs have a high impedance and are very easy to drive. This means that CMOS inputs do not draw much current or significantly affect the circuit driving them. Fan-out capabilities are also vastly improved over TTL, and some CMOS devices can drive five times as many other inputs as comparable TTL devices.

CMOS tends to be a more "forgiving" logic family than TTL. Minor power supply variations and noise do not bother CMOS to the extent they bother TTL. You can also forget about the de-glitching capacitors necessary in TTL work, and component layout is not as critical as with TTL (although leads should still be as short as practical in higher-frequency applications). CMOS is also less "fussy" than TTL about the purity and shape of clock pulses.

CMOS is not a perfect logic family, however. The operating speed of ordinary CMOS is inferior to TTL, although new high-speed versions of CMOS have closed this gap significantly. Far more serious is the susceptibility of CMOS devices to damage from static electricity. This is because the internal construction of CMOS devices involves several thin layers atop the silicon substrate; these layers form a "capacitor" that can be punctured by stray electric charges. The possibility of damage can be greatly reduced by careful handling and storage of CMOS devices and proper construction techniques; moreover, contemporary CMOS devices are improved over earlier ones in their resistance to damage from static charges. However, you will inevitably "zap" some CMOS devices along the way, particularly in dry weather (such as midwinter, during which static electricity charges can build up on clothing and carpets).

On balance, though, CMOS is a superior logic family to TTL and is preferable in most hobbyist and experimenter applications. Since the "action" in IC design has now shifted to CMOS, becoming skilled in using CMOS is vital for anyone in electronics. As mentioned in Chapter 4, there are currently CMOS equivalents for most popular TTL devices. These CMOS equivalents are indicated by "C" or "HC' after the first two digits of the TTL part number. Such "C" and "HC" devices are fully pin-for-pin equivalent to the standard TTL device and have identical functions and internal logic diagrams. However, they are completely CMOS and must be used and operated under the same conditions as CMOS. The availability of these CMOS equivalents to TTL ICs means that almost any requirement in digital logic can be met using CMOS. This wide array of available CMOS devices alone is a good reason for choosing CMOS. In fact, if you are not currently working with digital logic, your best bet would be to go with CMOS exclusively instead of starting with TTL.

In this chapter we will not examine those CMOS devices that are equivalents of TTL devices discussed in Chapter 4. Instead, we examine only those popular CMOS devices that have no equivalent in TTL. Table 5-1 gives a list of such "CMOS-only" devices. Information on internal diagrams, functions, pin connections, and logic tables given in Chapter 4 for TTL devices is equally valid for "74C" and "74HC" series devices as long as CMOS operating requirements are observed. In particular, the CMOS equivalents of popular TTL counters and shift registers will be very useful and are, in many cases, superior to similar devices available only in CMOS.

TABLE 5-1 Major CMOS IC Devices

Number	Description
4000	Dual three-input NOR gate with inverter
4001	Quad two-input NOR gate
4002	Dual four-input NOR gate
4009	Hex inverter
4011	Quad two-input NAND gate
4012	Dual four-input NAND gate
4013	Dual D flip-flop
4017	Decade counter/divider
4019	Quad two-input multiplexer
4020	14-stage binary counter
4021	PISO/SISO shift register
4022	Synchronous octal counter
4023	Triple three-input NAND gate
4024	Seven-stage binary counter
4025	Triple three-input NOR gate
4026	Decade counter/seven-segment LED driver
4027	Dual J-K flip-flop
4028	1-of-10 decoder
4029	Decade/hexadecimal up/down counter
4030	Quad two-input XOR gate
4042	Quad latch
4049	Hex inverting buffer
4050	Hex noninverting buffer

TABLE 5-1 (Continued)

Number	Description
4051	Analog multiplexer
4056	BCD-to-seven-segment decoder/LCD driver
4066	Quad bilaterial switch
4068	Eight-input NAND gate
4069	Low-power hex inverter
4070	Quad two-input XOR gate
4071	Quad two-input OR gate
4073	Triple three-input AND gate
4078	Eight-input NOR gate
4081	Quad two-input AND gate
4082	Dual four-input AND gate
4093	Quad Schmitt trigger
4097	Dual 1-of-8 analog switch
4192	Decade up/down counter
4193	Binary up/down counter
4502	Tri-state hex buffer
4503	Tri-state noninverting hex buffer
4510	Divide-by-10 up/down counter
4511	BCD-to-seven-segment latch/decoder
4512	Eight-channel data selector
4514	High-output 1-of-16 decoder
4515	Low-output 1-of-16 decoder
4516	Divide-by-16 up/down counter
4518	Dual divide-by-10 counter
4520	Dual divide-by-16 counter
4522	Decimal programmable counter
4526	Hexadecimal programmable counter
4527	Decimal rate multiplier
4528	Dual monostable multivibrator
4531	12-input parity generator
4532	Eight-level encoder
4539	Dual four-input data selector
4555	Dual 1-of-4 decoder
4556	Dual inverting 1-of-4 decoder
4584	Hex Schmitt trigger

VARIETIES OF CMOS DEVICES

Digital CMOS devices normally have part numbers consisting of four digits beginning with "4." The exception, as mentioned earlier, comes with the CMOS equivalents of popular TTL devices. For "original" CMOS digital ICs, there are two main groups. These are indicated by an A or B suffix to each part number. The meanings of these are as follows:

A. These are the earliest versions of various CMOS devices. As a rule, these have a slightly lower maximum supply voltage rating.

B. These are functionally identical to A series devices, but their outputs have internal buffers added. This makes the outputs of such devices more uniform. B series devices also usually have a slightly greater maximum supply voltage rating.

The addition of buffered outputs in the B series makes for more reliable and trouble-free circuits than A series devices. The difference is cost between the two versions is minor, so the B series is preferable for experimenting and most other applications. The A series is still preferable for a few uses, however, particularly in interfacing circuits.

USING CMOS

Preventing damage due to static electricity is the prime consideration in storing and using CMOS digital ICs. Fortunately, no extraordinary steps are necessary—common sense and a few simple precautions are all you need to observe.

The most obvious step is to avoid touching the pins of any CMOS device. Moreover, never apply an input signal to a CMOS device when the device is off. Also, drain any accumulated static charges from your body by touching a grounded object before handling a CMOS device.

Unused CMOS devices should be stored, *pins down,* on a conductive surface. Examples are an aluminum tray or surface covered with a sheet of aluminum foil. There are also special *conducting plastic foam* holders for CMOS devices which are excellent for storage; CMOS ICs will frequently come packaged in such foam. However, most plastic foam is *not* conducting, so do not store CMOS devices in plastic foam unless you are sure that it is the conductive type.

If you decide to make permanent versions of CMOS circuits, a battery-powered soldering iron must be used to solder any CMOS device's pins. Even better is to avoid soldering the CMOS device directly; IC sockets or wire wrapping are preferred.

Volt-ohmmeters and similar "multitesters" are often used to test other components in a circuit in which CMOS devices are used. In such cases, avoid touching any pins of the CMOS ICs with the tester's probes and make sure that voltage from the component or section under test does not flow to any of the pins of a CMOS device.

As mentioned earlier, CMOS devices can operate on +5 V although a higher voltage is preferred. Since CMOS consumes so little power, battery power supplies are more practical than with TTL. The circuit in Figure 5-1 supplies +9 V, which gives good performance with CMOS.

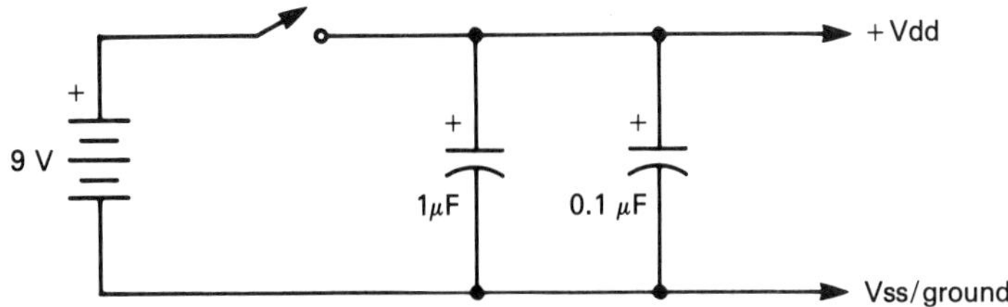

Figure 5-1 +9-V battery power supply for CMOS devices.

All CMOS inputs must go somewhere. Thus unused inputs should be set to high (Vdd) or low (Vss or ground) as appropriate. *The input voltage must not exceed Vdd.*

LEDs can be added to the outputs of CMOS devices much in the same manner as with TTL devices. A common practice, however, is to drive LEDs through buffers rather than directly. Figure 5-2 shows LEDs being used in conjunction with an inverter (NOT gate) such as that provided by the 4049 CMOS hex inverter. LEDs may be driven directly by the output of a B series CMOS device without the 1000-Ω current limiting resistor as long as the supply voltage of the device is less than +10 V.

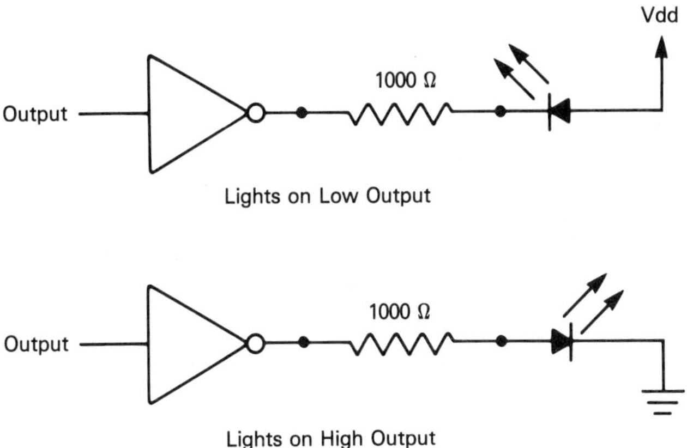

Figure 5-2 Using LEDs through inverting buffers with CMOS.

CMOS LOGIC GATES

CMOS equivalents are available for virtually every TTL logic gate IC, and as mentioned earlier, most of the logic gate circuits discussed and implemented using TTL in Chapter 4 can also be implemented using CMOS. Figure 5-3

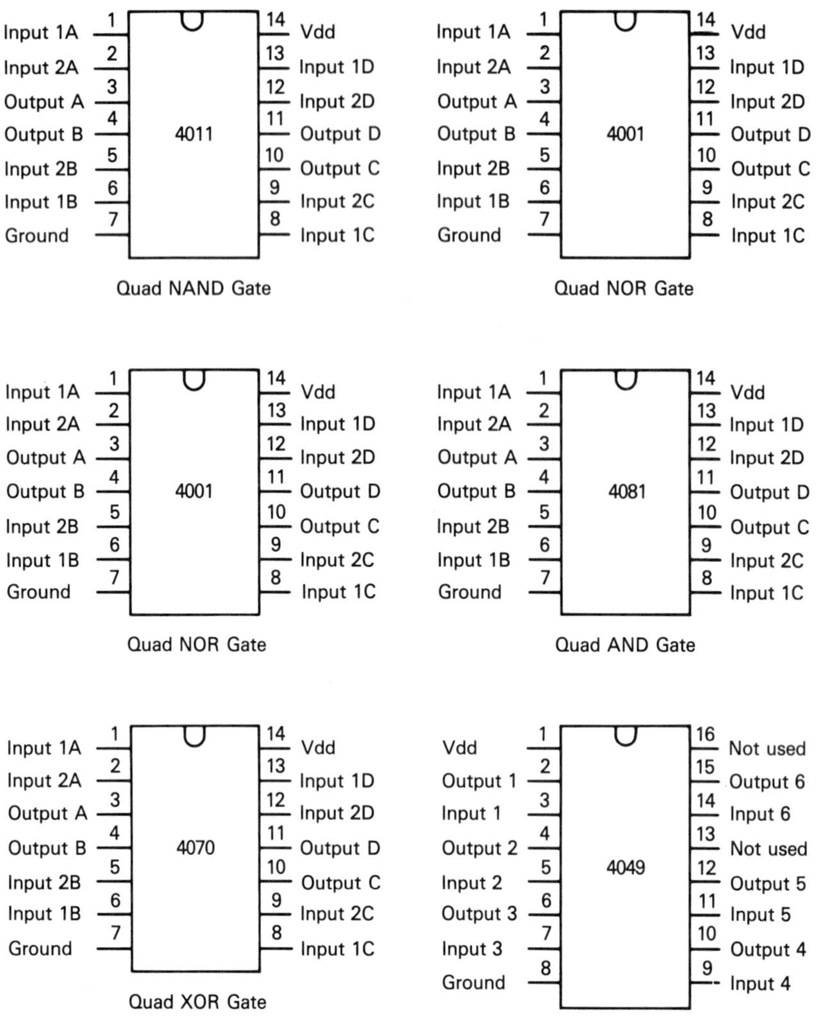

Figure 5–3 Pin connections for CMOS logic gate devices.

shows the pin connections and component numbers for the most popular CMOS logic gate devices. These can be used and combined to make multiple-input gates and new logic functions (such as XNOR) following the guidelines given in Chapter 4.

In addition to their similarities to their TTL equivalents, CMOS logic gate devices have several capabilities that make them superior in many applications to TTL. Figure 5–4 shows three circuits that can be implemented using the 4011 quad two-input NAND gate device. Like its TTL equivalent,

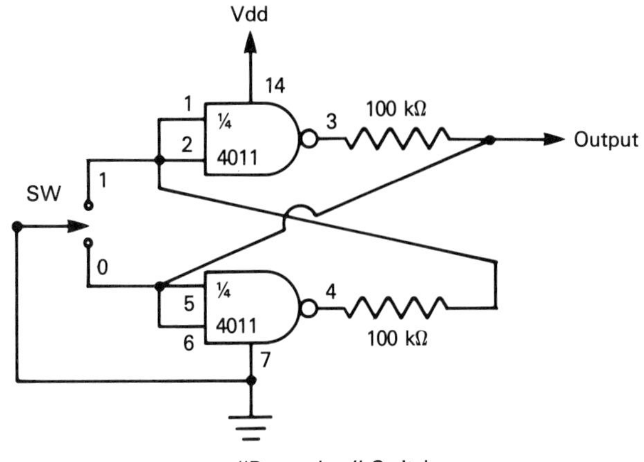

"Bounceless" Switch

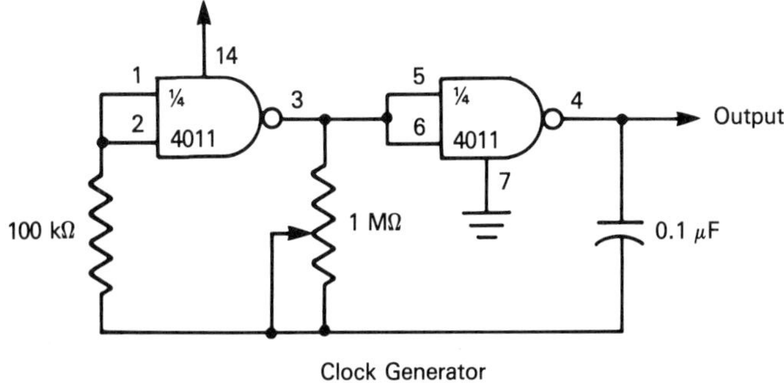

Clock Generator

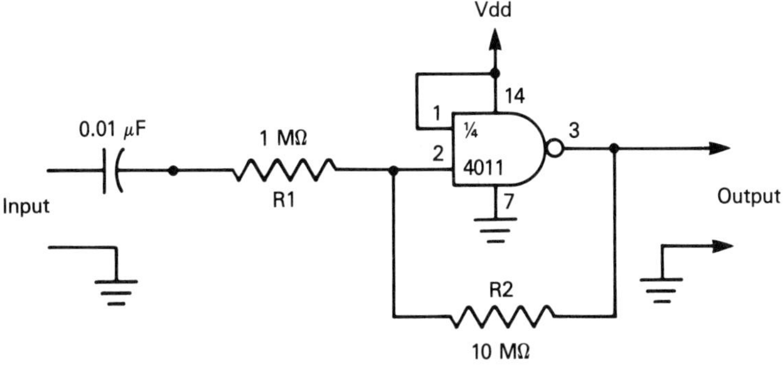

Linear Amplifier

Figure 5-4 Applications of the 4011 NAND-gate device.

the 7400, the 4011 is an extraordinarily versatile IC which can be used to implement any other logic function or circuit. The first circuit in Figure 5–4 is a *bounceless switch.* This circuit will emit a "pure" high or low logic state signal, depending on the setting of switch SW. This circuit should be very useful in experimenting with CMOS.

Another useful circuit is the *clock generator* shown in Figure 5–4. This circuit can be used to supply clock signals needed by many CMOS circuits; the output frequency can be adjusted by the I-MΩ potentiometer. This is another circuit for which you might want to build a permanent version if you plan to do much experimenting with CMOS devices.

The final circuit in Figure 5–4 illustrates how the 4011 is more versatile than the 7400, since the 4011 can be used in *linear amplifiers.* The circuit shown will amplify a voltage present at the input 10 times; the gain of this circuit is determined by R2/R1. In this case, the gain is 10 MΩ/1 MΩ, or 10. This is a *nondigital* use for a digital gate and is impossible to implement with the 7400 or other TTL logic. (Try it if you don't believe it.)

Two other uses for the 4011 are shown in Figure 5–5. The first circuit shows a *tone generator* driving an 8-Ω speaker to produce a 1000-Hz tone. If this circuit is "balky" in operation, add a resistor of approximately 1 MΩ between the 100-kΩ resistor and the inputs of the first NAND gate (at the point indicated by "A" in the schematic). While the output is used to drive a speaker, this circuit can also be used as a 1000-Hz clock signal generator.

Note that pins 1 and 2 are tied together to form an inverting buffer from the first NAND gate. It is possible to "split" pins 1 and 2 apart and leave one pin (such as pin 1) without an input from the rest of the circuit. An input could then be supplied from an external source; a high-level signal would then produce a tone or a stream of clock signals.

The second circuit shown in Figure 5–5 is an *LED flasher.* This circuit uses a buffer, formed by tying together the inputs of a NAND gate, for each LED. The frequency at which the LEDs flash is controlled by the two electrolytic capacitors. The values used, 33 μF, cause the LEDs to flash approximately once each second. This rate can be altered by changing the values of the capacitors. The values must be equal, and proper polarity must be observed.

Many of the NAND gates in Figures 5–4 and 5–5 were used as inverters. Those could be replaced by one of the inverters available on the 4049 hex inverter device. Figure 5–6 shows three circuits using sections of the 4049; these have been indicated by "A," "B," and "C." The first circuit is a clock generator similar to the one in Figure 5–4. The output frequency depends on the values of resistor R and capacitor C. The value of C can range from 0.01 to 10 μF. The output frequency is found by the formula

$$\text{output} = \frac{1}{1.4RC}$$

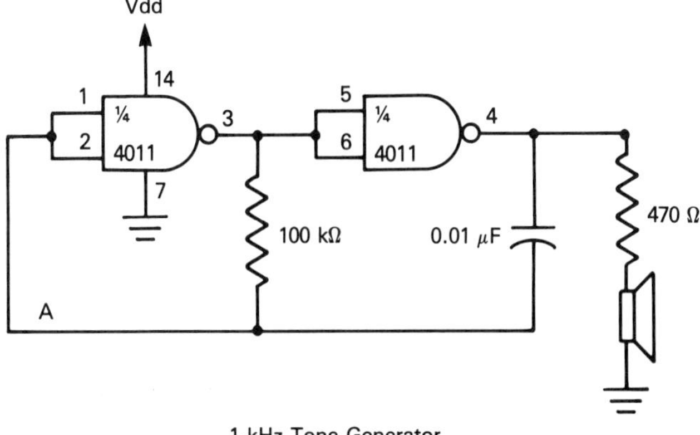

1-kHz Tone Generator

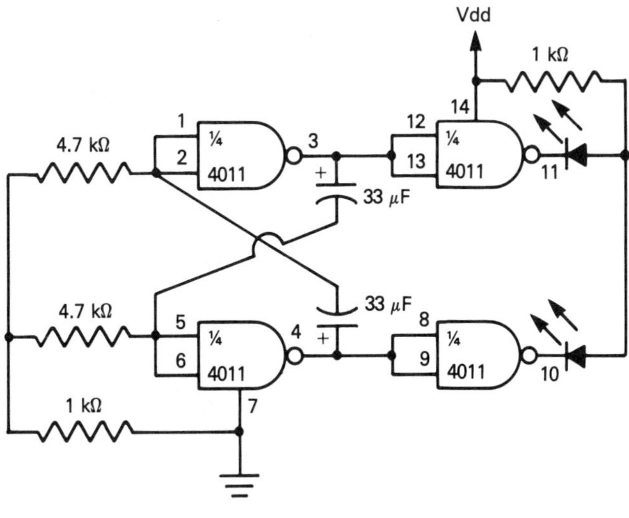

LED Flasher

Figure 5–5 Additional applications of the 4011.

The second circuit in Figure 5–6 is a *triangle and square-wave generator*. The output frequency is determined by the values of resistor R and capacitor C, and the same formula as given in the preceding paragraph is used to determine the output frequency. The final circuit in Figure 5–6 is a variation of the linear amplifier introduced in Figure 5–4; the gain of the circuit is again found by dividing R2 by R1. This is another circuit that cannot be implemented using TTL (the 7404 in this case).

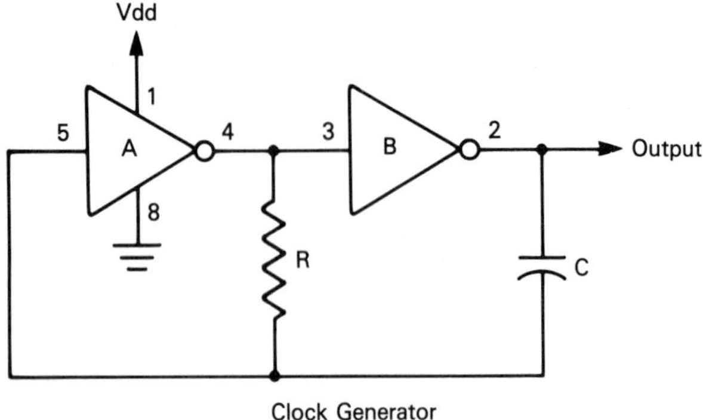

Clock Generator

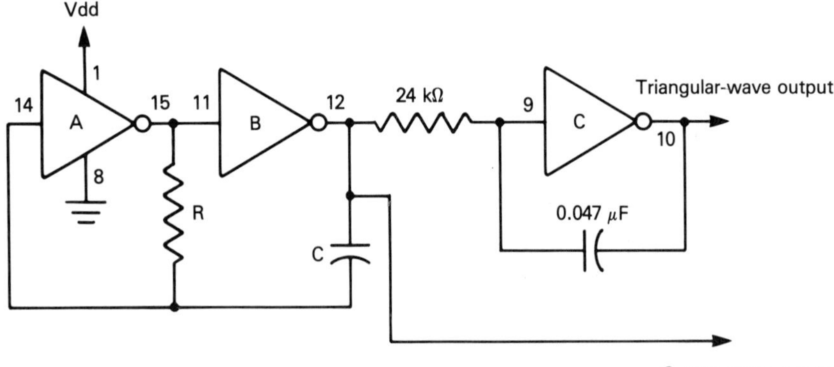

Triangular-/Square-wave Generator

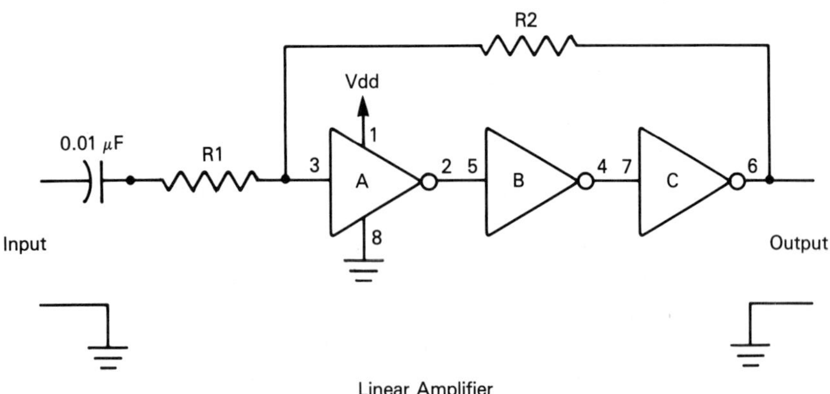

Linear Amplifier

Figure 5-6 Applications of the 4049 hex inverter device.

Other CMOS logic gates available are listed in Table 5-1. Note that several multiple-input devices are available; these should be used instead of implementing multiple-input logic gates from quad two-input devices.

THE 4047 ASTABLE/MONOSTABLE MULTIVIBRATOR

One interesting CMOS device is the 4047, whose pin connections are shown in Figure 5-7. Depending on how this device is used, it can function as either an astable or monstable multivibrator. When used as an astable multivibrator, it has an astable output together with a pair of complementary outputs, Q and $\overline{Q}$, whose frequencies are half that of the astable output. When used as a monostable multivibrator, it may be either positively or negatively triggered.

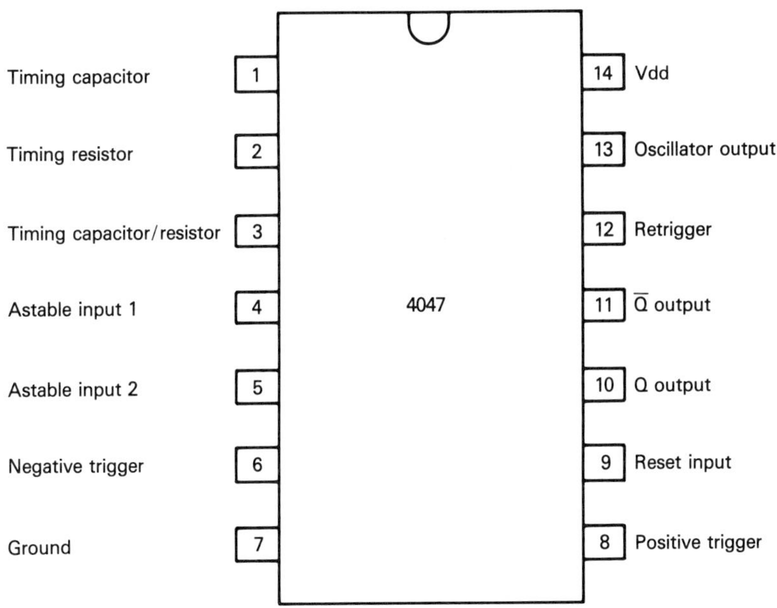

Figure 5-7 Pin connections of the 4047 monostable/astable multivibrator.

Figure 5-8 shows the three ways the 4047 may be configured. In all three circuits, the output frequency is controlled by the timing capacitor and resistor attached to pins 1, 2, and 3 of the 4047. In each case, the length of the output pulse increases as the values of the timing resistor and capacitor increase. The values shown for the astable multivibrator will produce an approximately 60-Hz output. When used as either monostable multivibrator, the width of the output pulse is found by the formula

$$\text{length} = 2.5RC$$

in which the pulse length is measured in milliseconds, resistance in kilohms, and capcitance in microfarads. For all three circuits, the timing resistor may range from 10 kΩ to 4.7 MΩ and the timing capacitor from 0.001 to 0.1 μF.

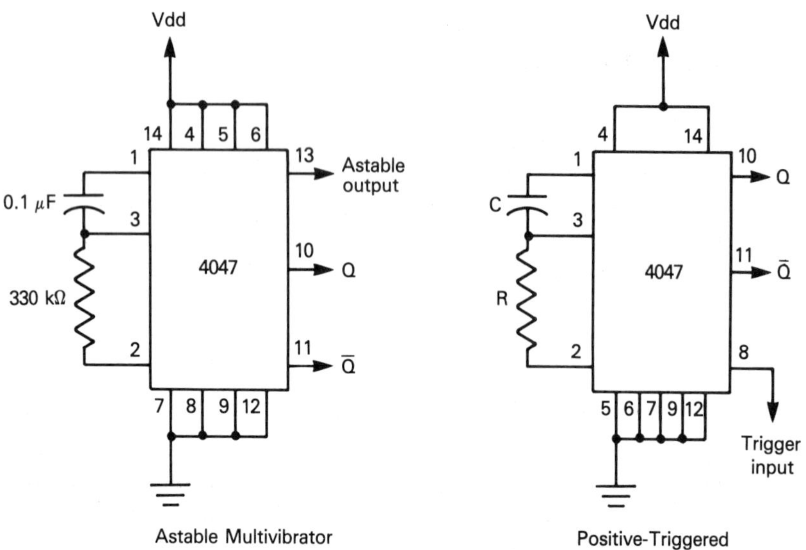

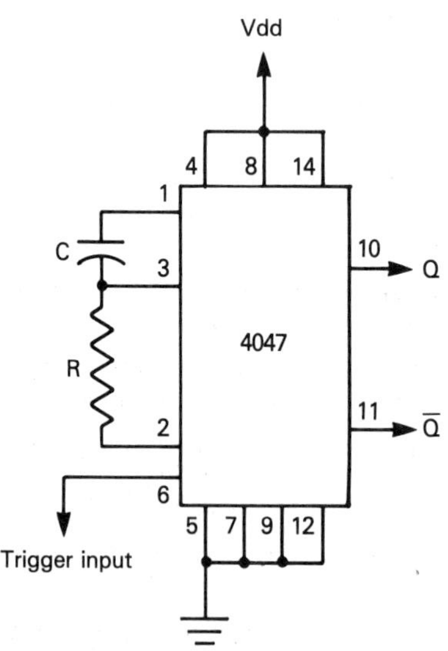

Figure 5-8 Applications of the 4047 device.

CMOS FLIP-FLOP DEVICES

Several flip-flop devices are available in CMOS. One that is a "CMOS exclusive" is the 4013 dual D flip-flop IC. Figure 5–9 shows the pin connections for this device along with two counter/divider circuits using it. (Note that the terms "set" for "preset" and "reset" for "clear" are often used with CMOS flip-flops to help distinguish them from TTL flip-flops.)

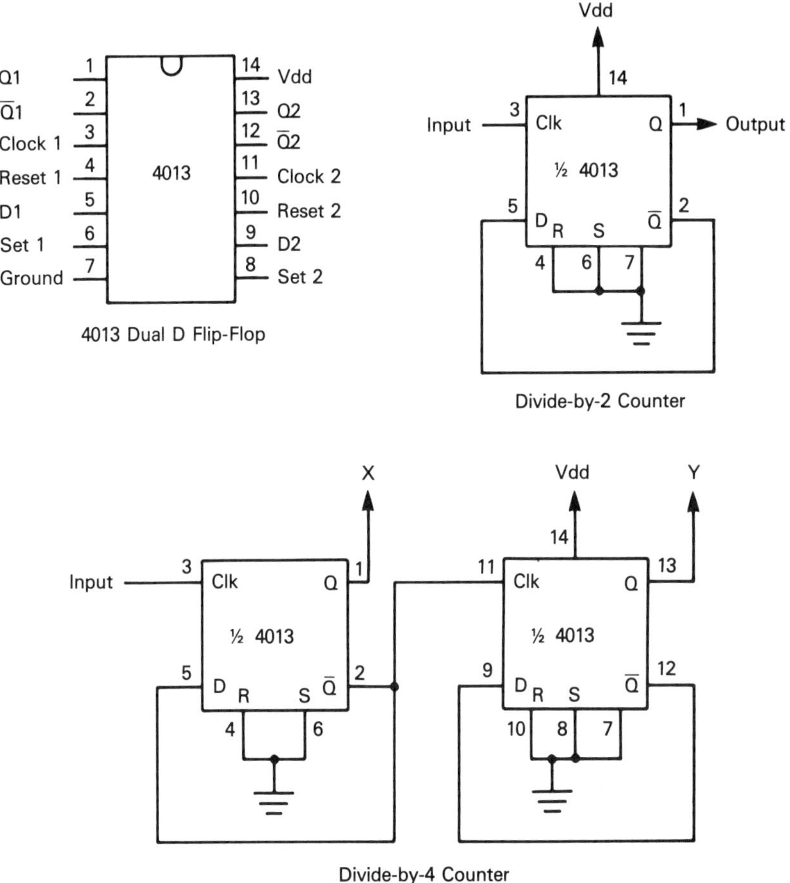

Figure 5–9 Pin connections and applications of the 4013 dual D flip-flop device.

When the set and reset inputs of either flip-flop are held to ground, the outputs at Q and $\overline{Q}$ are controlled by the input at D and the clock input. Like TTL D flip-flops, the 4013 is edge clocked and the *positive* edge of a transition from a low to a high logic state at the clock input allows D to control Q and $\overline{Q}$. The logic state at Q is the same as the state at D on each positive-edge transition.

The set and reset inputs can also be used to control each flip-flop directly. A high set input will make Q high and $\overline{Q}$ low, while a high reset input will make Q low and $\overline{Q}$ high. Applying a high-level input to both the set and rest inputs will make both Q and $\overline{Q}$ high, which is not a permitted condition under the rules of logic for a flip-flop. The *last* reset or set input to go low will control the state of Q and $\overline{Q}$. Figure 5–9 shows how the 4013 flip-flops can be chained together to form counter/divider circuits, similar to those shown in Figures 4–34 and 4–35.

The 4027 is a dual J-K flip-flop. Figure 5–10 shows its pin connections and two counter/divider applications. The following is a logic table for each

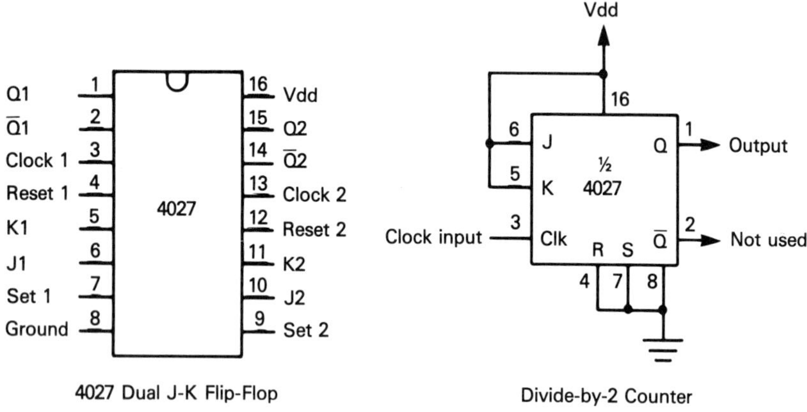

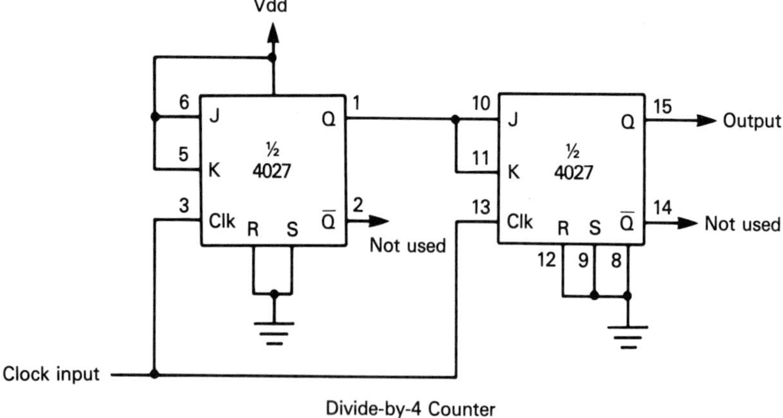

Figure 5–10 Pin connections and applications of the 4027 dual J-K flip-flop device.

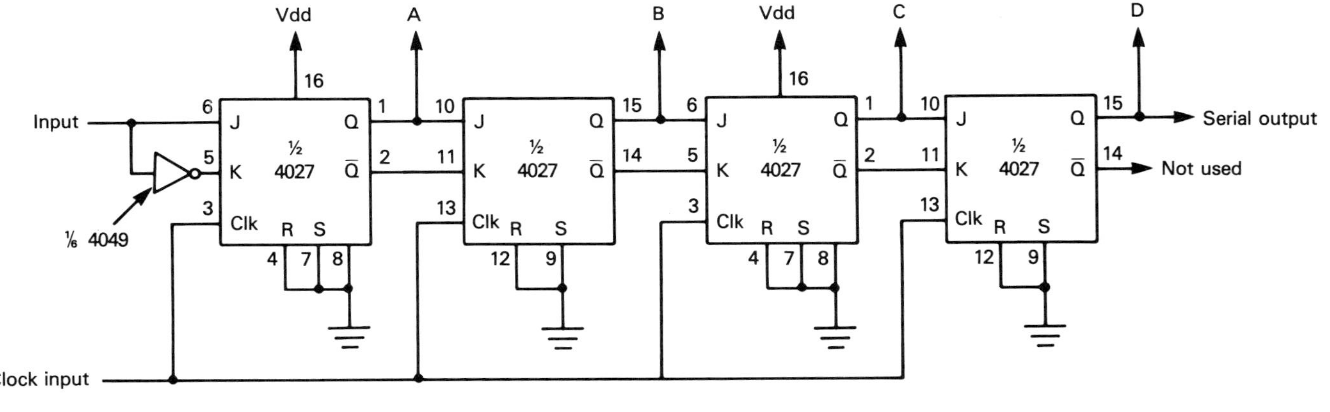

Figure 5–11 Four-bit SISO/SIPO shift register using the 4027.

flip-flop, with "X" indicating the state of the signal is irrelevant and "E" indicating that a clock signal is present:

Set	Reset	Clk	J	K	Q	$\overline{Q}$
H	H	X	X	X	Not allowed	
H	L	X	X	X	H	L
L	H	X	X	X	L	H
L	L	E	L	L	No change	
L	L	E	H	L	H	L
L	L	E	L	H	L	H
L	L	E	H	H	Outputs toggle in response to incoming clock pulses	

There is a significant difference between this J-K flip-flop and those discussed in Chapter 4, in that the 4027 uses positive-edge clocking rather than level clocking. This means that the J and K inputs can still accept data between clock pulses; the state of the J and K inputs when clocking takes place determines the outputs. The set and reset inputs override the J, K, and clock inputs.

Figure 5–10 shows the 4027 used as divide-by-2 and divide-by-4 counter/dividers; these operate in a similar manner to their TTL equivalent. Additional flip-flops may be chained together for a higher count. Two 4027 devices are used in Figure 5–11; this circuit is very similar to the one shown in Figure 4–31. The serial input is applied to the J input of the first flip-flop and through an inverter (such as one section of a 4049 or similar) to the K input. This circuit can be a serial input, serial output (SISO), or serial input, parallel output (SIPO) shift register; parallel output is available at outputs A, B, C, and D, while a serial output can be obtained from the Q output of the last flip-flop (which also provides the D parallel output). As with shift registers implemented using TTL flip-flops, each clock pulse causes each input bit to the shifted toward the right.

CMOS COUNTERS AND SHIFT REGISTERS

Counter and shift register ICs are available in CMOS. These are used in a similar fashion to their TTL counterparts. Figure 5–12 shows the pin connections and two applications of the 4017 decade counter/divider IC. This is a type of counter known as a *Johnson Counter.* This means that the input signal is applied simultaneously to all the internal flip-flop circuits and does not "ripple" down through the chain of flip-flops. As such, the 4017 is a *synchronous* counter device. It can provide an output that is one-tenth of

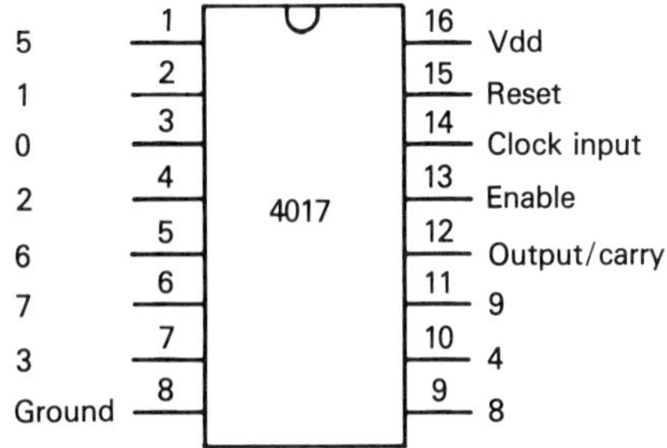

4017 Decade Counter/Divider

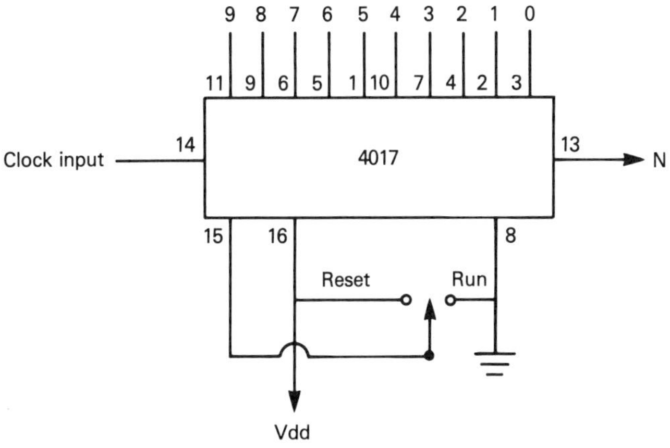

Count to N and Halt

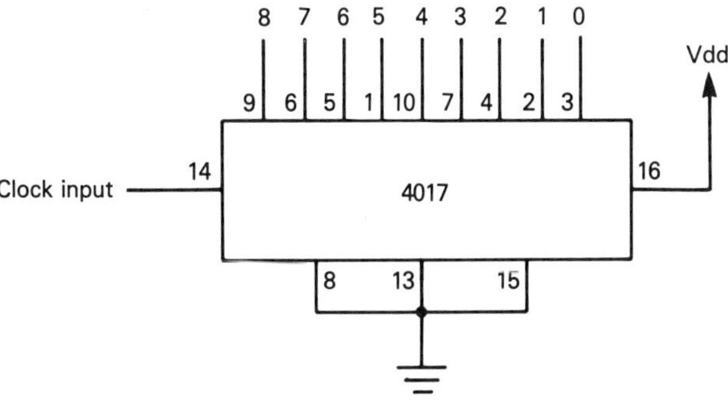

Divide-by-9 and Recycle

Figure 5–12 Pin connections and applications of the 4017 decade divider/counter.

the clock input frequency from pin 12 or a "decoded" output at pins 1 through 7 and pins 9 to 11. In the latter condition, the appropriate output will go high to indicate the count, while the others remain low.

For normal counting operation, the reset and enable inputs (pins 15 and 13) must both be low. The count will advance on each positive edge of the clock input at pin 14. A high-level input to the reset input will set the counter back to zero; when this happens, the output/carry output (pin 12) and the 0 output (pin 3) are both high, while the other outputs are all low. A high input to the enable input will "freeze" the count until the enable input goes low again.

The two applications of the 4017 shown in Figure 5-12 are relatively simple. In the count-to-N-and-halt, the counter will continue to count as long as the switch is set to the "run" position and the signal at the enable input is low. Setting the switch to reset will clear the counter and allow counting from zero again. In the divide-by-9 counter, the clock input is divided into

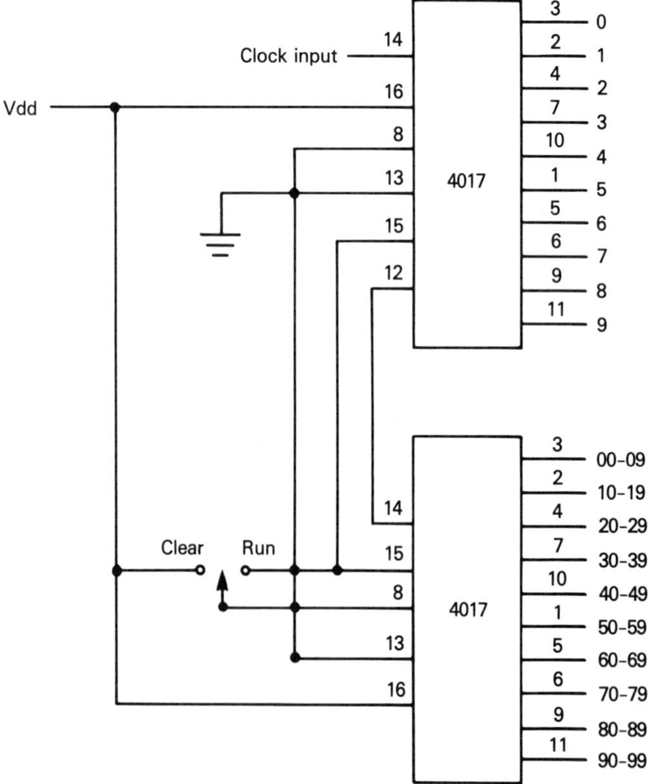

Figure 5-13 The 0-to-99 counter using cascaded 4017 devices.

nine outputs, after which the circuit automatically recycles back to zero. In both of these circuits, a serial output is available at pin 12 if desired.

Pin 12 is also used as a "carry" output if two or more 4017 devices are cascaded to count higher. Figure 5–13 shows a 0-to-99 counter used in such a manner; additional counters may be cascaded by connecting pins 12 and 14 as shown. Pin 12 can also be used to obtain a divide-by-100 serial output if needed.

A popular CMOS shift register is the 4021, which may be used as either a SISO or a PISO device. Figure 5–14 shows the pin connections and a typical application of this device. When used as a SISO device, the load input (pin 9) should be set to low. Data input to pin 11 is shifted internally on each positive clock edge. *The first output signal does not appear until after six*

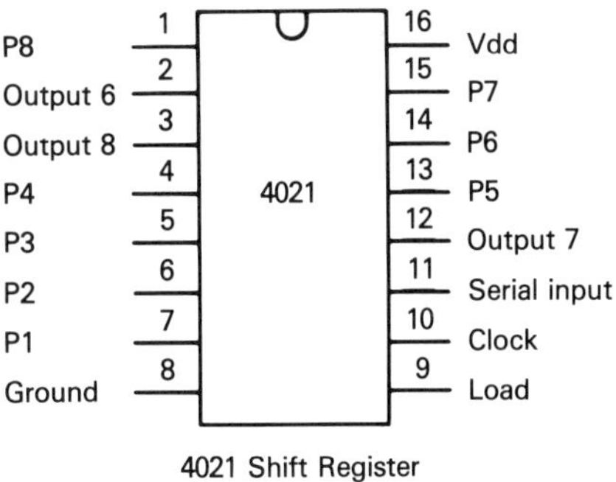

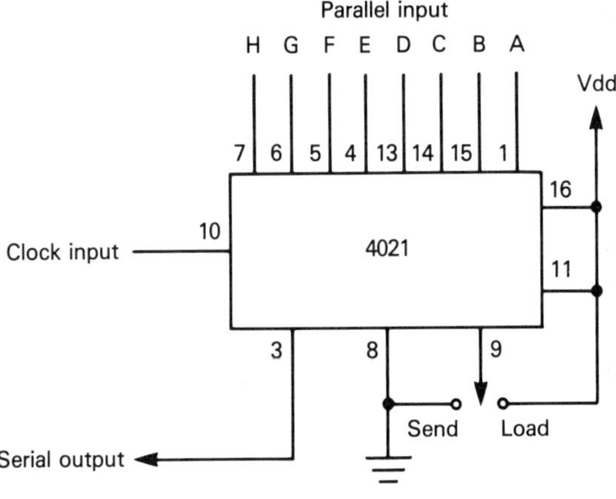

Figure 5–14 Pin connections of 4021 shift register and application in parallel-to-serial converter.

clock pulses. After six clock signals, an output appears at output 6. After seven clock signals, it appears at output 7. Finally, after eight o'clock signals it appears at output 8. This is why the 4021 is known variously as a *six-, seven-,* or *eight-stage shift register.*

To input data in parallel form, the inputs labeled P1 through P8 are used. To "load" these inputs into the shift register, the load input (pin 9) is briefly set to high and then immediately returned to low.

One common application for this IC is parallel-to-serial conversion, as shown in Figure 5–14. Parallel input is loaded into the shift register by setting the switch to the "load" position; the input is then converted to serial form and output at pin 3 by resetting the switch to the "send" position.

Perhaps the most versatile shift register device available is the 4034, whose pin connections are shown in Figure 5–15 (note that it comes in a 24-pin DIP). This is known as a *universal shift register.* It is capable of handling both parallel and serial data and operating in asynchronous or synchronous modes.

The 4034 has 16 pins divided into two sets, labeled A1 through A8 and B1 through B8. These two sets can be used for either input or output, with

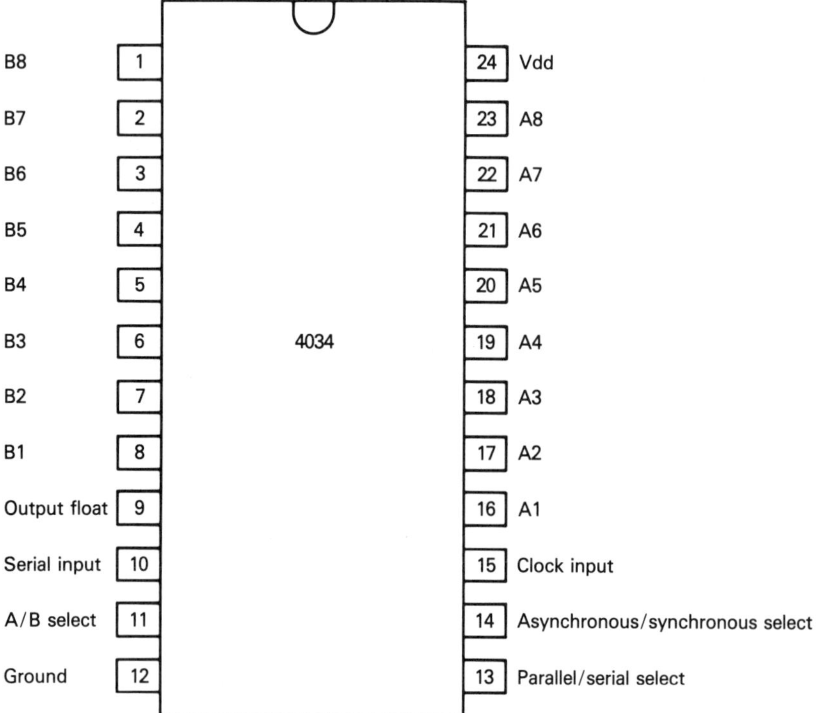

Figure 5–15 Pin connections of the 4034 universal shift register.

their action controlled by the A/B select input (pin 11). When a high-level input is present at pin 11, the A pins are inputs while the B pins are outputs. If the A/B select input is low, the opposite is true; the B pins are inputs while the A pins are outputs.

The asynchronous/synchronous select input determines operation of the device. If this input is high, data shifting takes place immediately (or asynchronously). If it is low, data shifting must wait until the next positive edge of a clock signal. The parallel/serial select input causes parallel operation if it is high; parallel operation may be synchronous or asynchronous. In parallel, data are moved between the sets of A and B pins. Data may also be entered serially at the serial input (pin 11) if the signal at the parallel/serial select is high. For serial operation, the signal at the asynchronous/synchronous select input must be low; serial input is shifted one position at each positive clock edge. Finally, the output float input permits a delay in output appearing at the A pins after it has been input at the B pins. If the output float input is low (and the input at A/B select is also low), data input at the B pins will "float" internally until the output float signal is made high; the output will then appear at the A pins.

THE 4093 QUAD TWO-INPUT SCHMITT TRIGGER

One popular CMOS Schmitt trigger IC is the 4093 quad two-input NAND Schmitt trigger device. Each NAND gate functions according to the same logic as an ordinary NAND gate (such as the 4011) but has been optimized for use as a Schmitt trigger. It is very similar to the 74132 TTL device covered in Chapter 4. The action of each of the four Schmitt triggers is similar to that of those discussed in Chapter 4, but the performance is better due to the greater noise immunity of CMOS. If only one input is needed, the two inputs can be connected. Figure 5-16 shows the pin connections for the 4093 and two applications circuits.

The *triggered astable multivibrator* circuit shown in Figure 5-16 will emit a 1-kHz square-wave output when a high-level input is present at the trigger input. The output of this circuit can be used to drive a speaker, making this circuit useful as an alarm or audible indicator of a certain condition.

Given CMOS's greater immunity to noise, one of the most common uses for the 4093 is to extract data from noisy signals. The *signal conditioner* circuit will give an output that is much "cleaner" (although inverted) than the input signal. This circuit is especially useful if input is being taken from TTL devices. Random variations in the input signal and transients are "swallowed" by the 4093 section, resulting in an output with a sharply defined rise-and-fall pattern.

If ordinary Schmitt triggers with a single input can be used, the 4584 has six inverter-type Schmitt triggers in one package.

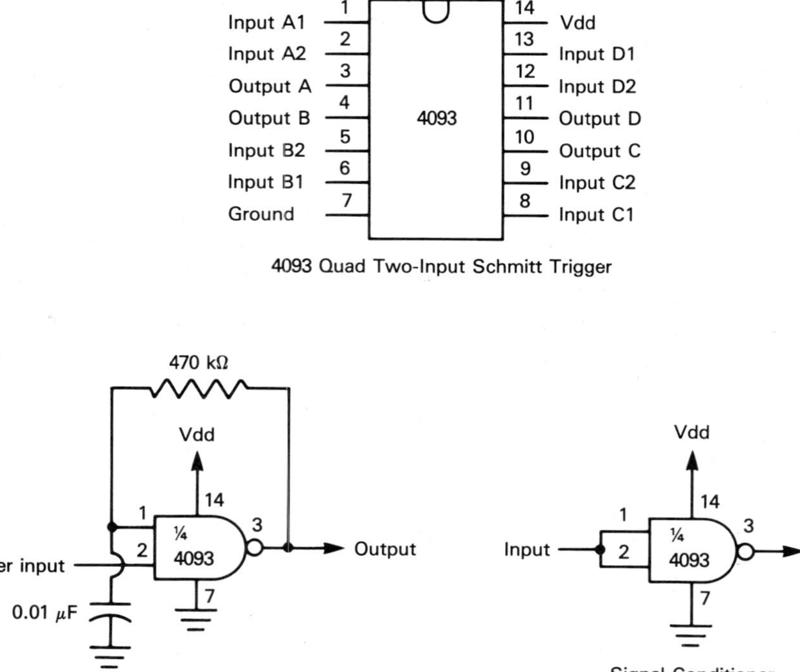

Figure 5–16 Pin connections and applications of the 4093 quad-Schmitt-trigger device.

THE 4066 QUAD BILATERAL SWITCH

The 4066 is an interesting device containing four independent switches in a single package, as shown in Figure 5–17. The switches may be used with either digital or analog signals; the switches can be used separately or in combinations. The various input/output (I/O) pins are reversible and can be used for either input or output. The switches are controlled by signals or voltages applied to the pins labeled A, B, C, and D.

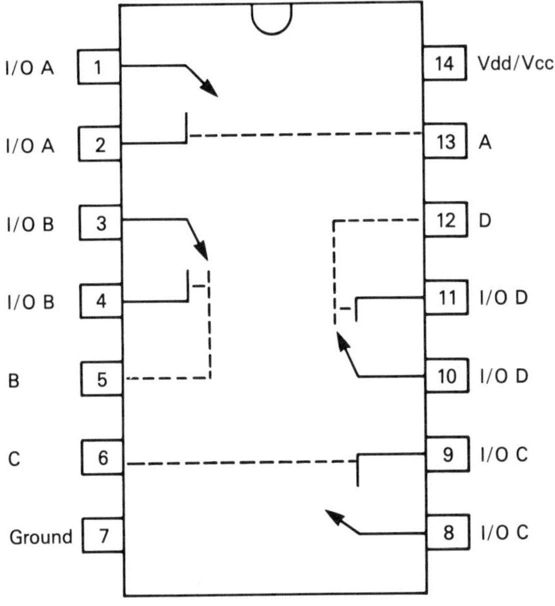

Figure 5–17 Pin connections and internal diagram for 4066 quad-bilateral-switch device.

Any of the switches is off (i.e., has a very high input impedance) when the voltage applied to it equals the voltage at pin 7. When the voltage applied equals that at pin 14, the switch is on and has a very low resistance (usually less than 250 Ω). All voltages through the switches must not exceed the voltage at pin 14 nor drop below that at pin 7. The usual convention for the analog mode of operation is to have a voltage of +5 V at pin 14 and a voltage of −5 V at pin 7. In digital operation, normal CMOS Vdd and ground voltage conventions are followed.

Figure 5–18 shows two uses for the 4066. The *data gate* is used to control the flow of information into a circuit (such as from a data bus) by means of a control voltage applied simultaneously to pins 5, 6, 12, and 13. A low signal applied to those pins will "close" the gate and prevent the signals at

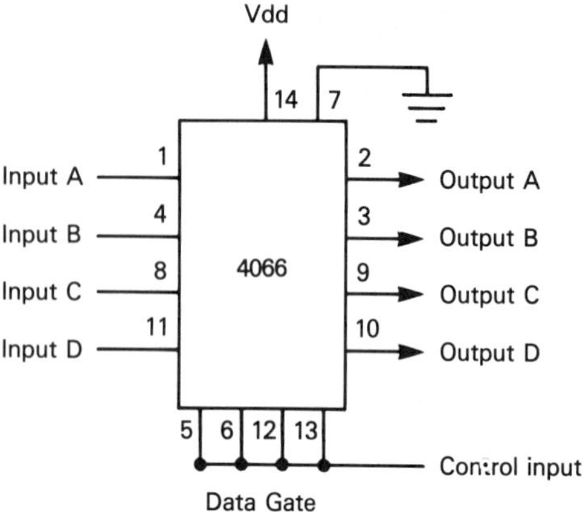

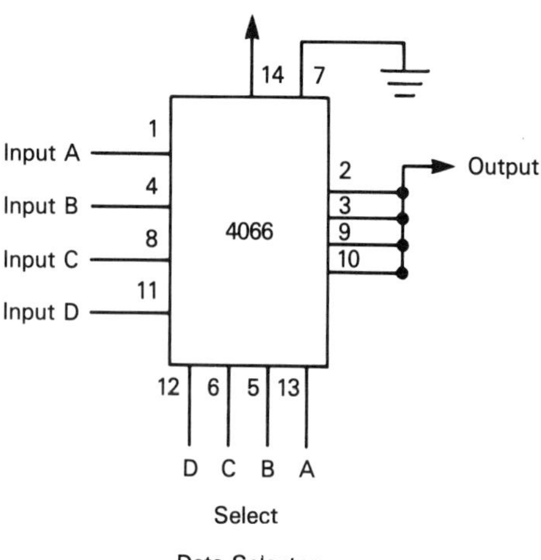

Figure 5-18 Applications of the 4066.

the inputs from appearing at the outputs. A high signal "opens" the gate and lets the same signal appear at both input and output.

The *data selector* circuit can be used to select one desired input. A high signal is applied to the select input corresponding to the desired input (A, B, C, or D) while the other select inputs are held low. The desired input will then appear at the output. This particular circuit can be modified by "disconnecting" the outputs from a common point (as was done in the data gate circuit) so that two or more desired outputs can be obtained simultaneously.

The 4066 can be a remarkably useful device. You can use it almost anywhere you would use a mechanical switch. For example, you can use it to switch enable signals on or off in digital circuits or to select different feedback resistors in op amp circuits for variable gain. Since it can be used in linear circuits as a linear device, it eliminates the need for complicated interfacing. The maximum switching speed increases with supply voltage; typical values are 10 MHz at +10 V and 5 MHz at +5 V.

THE 4051 1-OF-8 DIGITAL/LINEAR SWITCH

The 4051 is another CMOS device that may be used as a linear or digital device. Its most common application is as a *multiplexer* or *demultiplexer.* A demultiplexer selects one of several inputs and "forwards" it to its output. A multiplexer does the opposite: it takes an input and sends it to one of several possible outputs. The selection of inputs or outputs is controlled by address input signals to the multiplexer or demultiplexer. Both circuits may be implemented from AND and OR logic gates, but are easier to use in integrated form.

Figure 5–19 shows the pin connections of the 4051 and its use in multiplexer and demultiplexer circuits. Like the 4066, it contains internal switches that can be used with either linear or digital signals. However, there is a major difference between the 4051 and 4066. Each of the four switches in the 4066 had its own pair of input/outputs pins. By contrast, the eight switches of the 4051 have a pin that can be used for input or output (identified as 0 through 7 in Figure 5–18) but must share one common input/output point (pin 3) with each other. Thus it is possible to distribute one input at pin 3 to the other switches' pins or to direct input from the other eight pins to pin 3 for output, but not to have the independent input or output possible with the four switches of the 4066. This is why the 4051 is described as a "1-of-8" switch. Thus one of the eight different inputs may be selected to produce an output at pin 3 while a single input can appear at any of the eight outputs.

Switches are selected by the A, B, and C select inputs (pins 9, 10, and 11). The inputs to the A, B, or C select switches must be binary words, with A = 1, B = 2, and C = 4. The inhibit input (pin 6) opens all switches and disables the device when a high signal is applied to it but permits normal operation when low. The supply voltage for linear operation is +5 V; pin 7 must be connected to −5 V as well. Normal CMOS requirements apply when used in digital operation. The circuits in Figure 5–18 are set up for digital operation.

As was the case with TTL circuits, CMOS devices invite exploration and experimentation. Additional information on the devices discussed in this chapter and other CMOS devices can be obtained from the manufacturer's data sheets and applications notes. Observe normal precautions when using CMOS and—as with TTL—the worst thing that can happen is a nonfunctioning circuit.

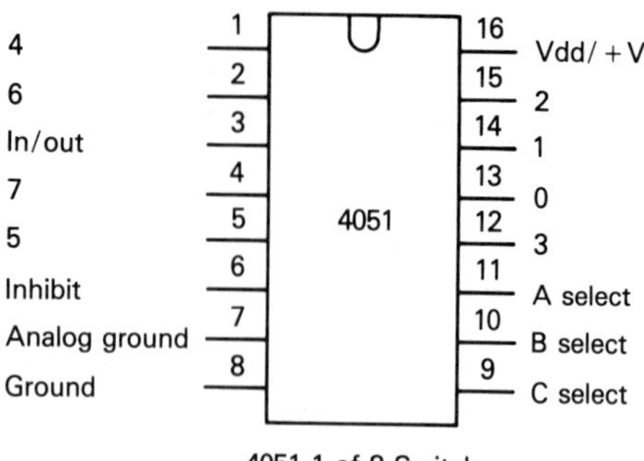

4051 1-of-8 Switch

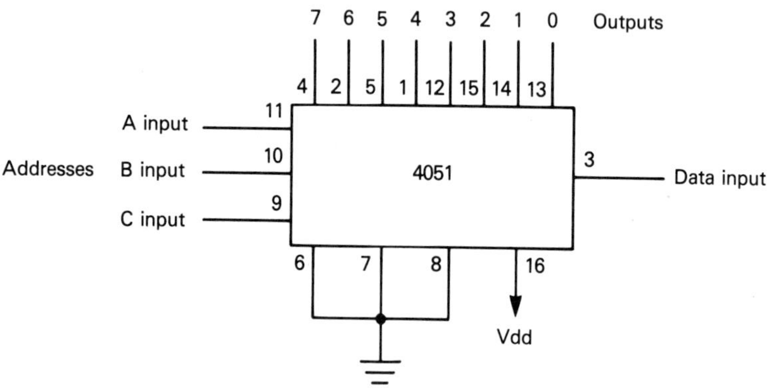

1-of-8 Multiplexer

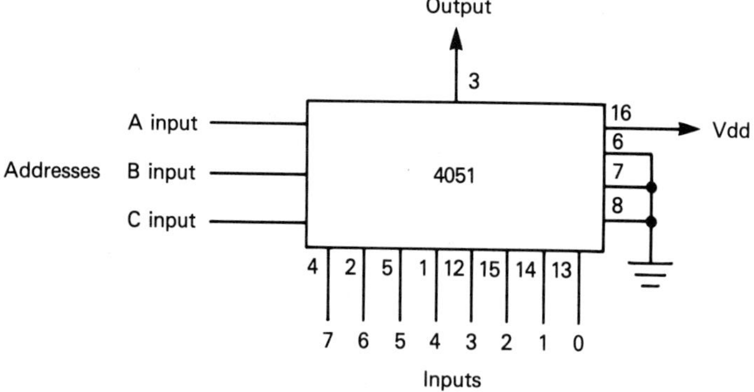

1-of-8 Demultiplexer

Figure 5-19 Pin connections and applications of the 4051 1-of-8 switch.

chapter 6

TYING IT ALL TOGETHER

The circuits covered so far in this book have been interesting but in most cases are not too useful by themselves. Most circuits have to be used in conjunction with other circuits, and that often involves some interfacing and troubleshooting, the subjects of this chapter.

INTERFACING TTL AND CMOS

The best way to handle the problem of using both CMOS and TTL logic together in the same circuit is—*don't*! Since most popular TTL devices now have direct CMOS equivalents (the "C" and "HC" part numbers), there is often no need to mix the two logic families in the same circuit. You can save yourself a lot of grief if you stick with just one logic family, whether TTL or CMOS, in a circuit.

However, you may be faced with situations in which you have no choice but to mix CMOS and TTL logic in the same circuit design. Then some sort of interfacing is necessary so the different logic families can "talk" to each other. The most obvious problem is the different supply voltage requirements for the two families; if they are equal, matters are greatly simplified.

Figure 6–1 shows two approaches to interfacing TTL to CMOS. The first circuit is for those situations in which a common +5-V power supply is being used. In such cases, TTL can drive CMOS through a *pull-up resistor* that delivers a higher input voltage to the CMOS device. The pull-up resistor can be in the range 1 to 3.9 kΩ with values at the higher end of this range preferable for LS versions of TTL. This circuit can allow a TTL device to drive as many CMOS inputs as you are likely to encounter in real-world situations.

Since many of the advantages of CMOS are lost when operating at +5 V, it is usually preferable to run CMOS at a higher voltage. The second circuit in Figure 6–1 was designed to boost the output of a TTL circuit so that it can drive a CMOS circuit operating at +12 to 15 V. This circuit is known as a *common-base level shifter* after the configuration of the 2N2222 transistor. This circuit preserves the greater noise immunity of CMOS.

If CMOS is to drive TTL, and a common +5-V supply voltage is used, a single CMOS output can directly drive one TTL input without any interfacing. The 4049, 4050, and 4502 device outputs can drive up to two TTL inputs. If additional devices need to be driven, an "input expander" composed of CMOS logic gates should be used.

When different power supply voltages are used, CMOS-to-TTL interfacing must be through a transistor. The circuit in Figure 6–2 uses a MOSFET transistor as a buffer between the two logic families. The maximum Vdd for this circuit is +15 V.

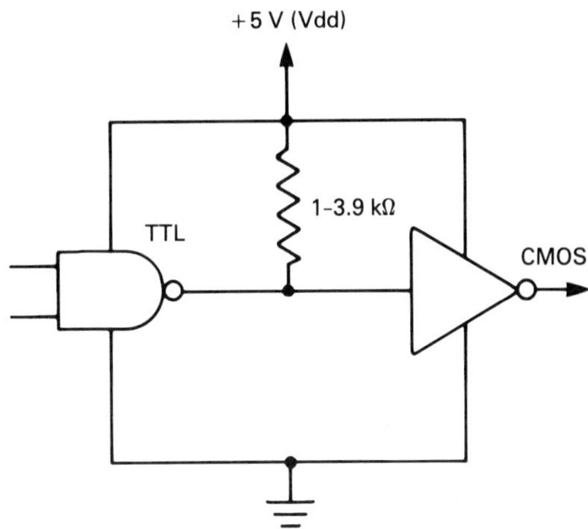

TTL-to-CMOS Interfacing (Common Supply Voltage)

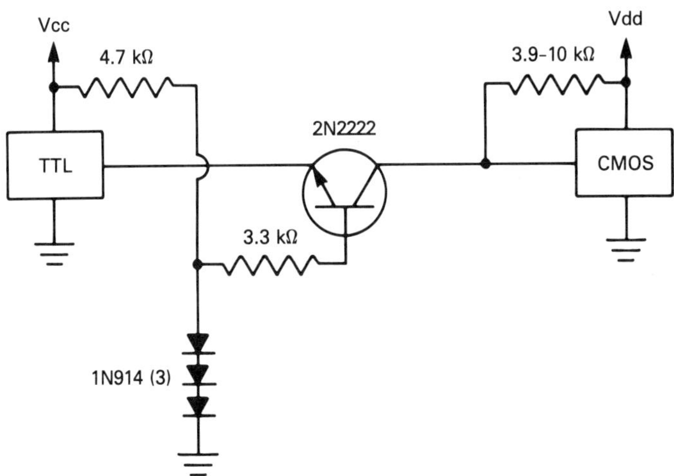

TTL-to-CMOS Interfacing (Different Supply Voltages)

Figure 6–1 TTL-to-CMOS interfacing circuits.

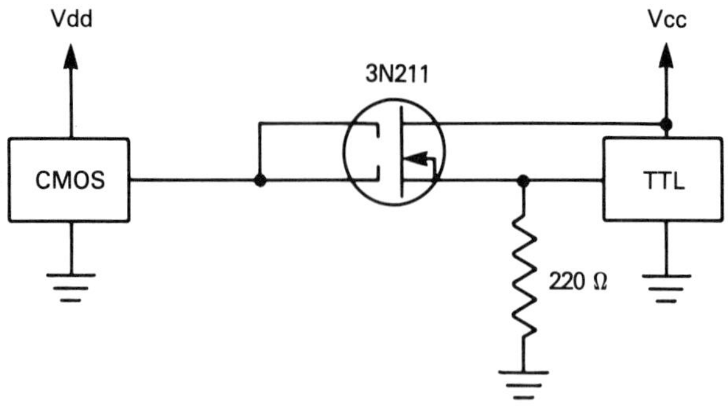

Figure 6–2 CMOS-to-TTL interfacing for different supply voltages.

INTERFACING WITH OTHER COMPONENTS AND DEVICES

Driving external loads such as relays is generally easier with TTL than CMOS, since TTL can "sink" much more current than CMOS. Figure 6–3 shows a generalized external interfacing circuit using a PNP transistor to carry the current necessary to drive an external load. The exact PNP transistor necessary must be selected on the basis of the current requirements of the load.

Special low-current relays operating from +5 V have been developed for using with TTL devices. These relays come in DIPs similar to ICs and can fit onto solderless breadboards. Figure 6–4 illustrates how to use such a device with TTL. The 1N914 diode is used to prevent a voltage "spike" from the relay (developed by the changing magnetic fields when the relay opens and closes) causing damage to the TTL device.

Driving CMOS and TTL from an op amp is a relatively simple matter, as shown in Figure 6–5. To drive TTL from an op amp, a pull-up resistor in the range 3.3 to 4.7 kΩ will increase the output of an op amp to a level

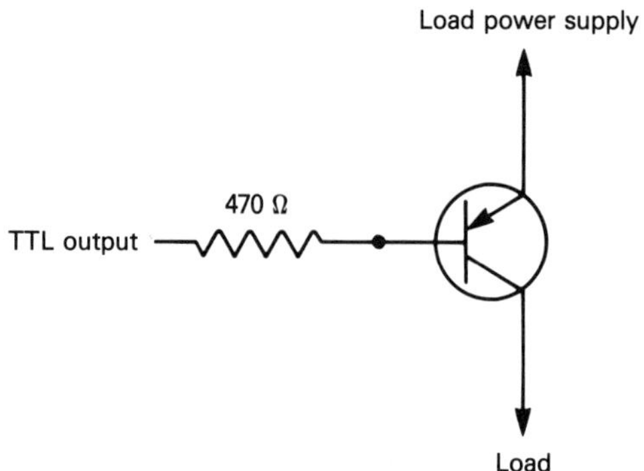

Figure 6–3 Driving external loads from TTL outputs.

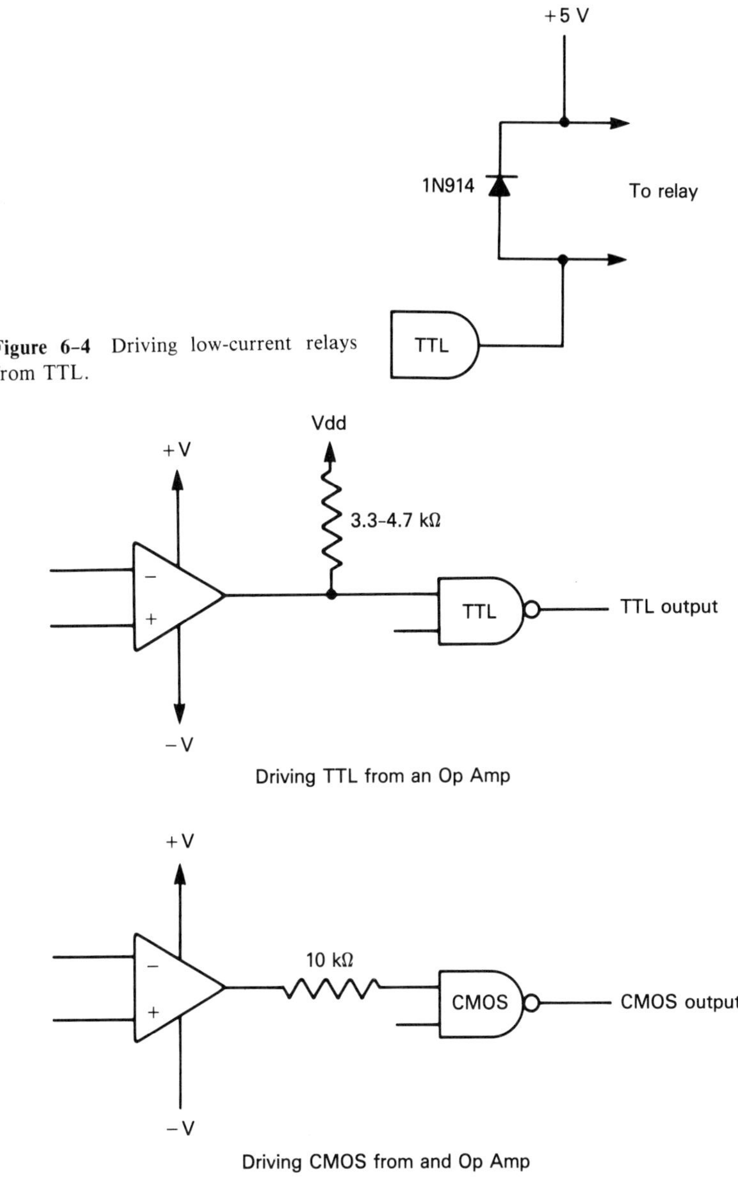

Figure 6–4 Driving low-current relays from TTL.

Figure 6–5 Driving TTL and CMOS from operational amplifiers.

sufficient to serve as a TTL input. Driving CMOS is even easier. A resistor of approximately 10 kΩ is inserted in series between the op amp output and the CMOS device input; the resistor serves to limit the amount of current reaching the CMOS input.

As mentioned earlier in this book, several linear devices, such as op amps and timers, which have been fabricated using CMOS technologies are available. Most of these have outputs that are directly compatible with CMOS and TTL logic. (For example, the TLC555 timer is identical to the "standard" 555 timer in functions and applications, but its outputs can directly drive CMOS and TTL devices.) In addition, some other linear devices constructed using bipolar technologies are available whose outputs are compatible with digital logic. The data sheet and applications notes for various devices will indicate if they are CMOS/TTL compatible, and usually this feature will be prominently featured in advertising and catalog descriptions of the device. Selecting such components to use with digital logic ICs is a wise choice. There will be some additional cost, but the savings in time and headaches will be worth it!

DEBUGGING AND TROUBLESHOOTING

It is often difficult to get a circuit working right on your first attempt, especially if you are just starting to work with ICs. Problems fall into two broad categories: (1) the circuit does not work at all, or (2) the circuit does not work correctly.

There is one rule that will save you a lot of time when troubleshooting and debugging a circuit: *Blame yourself first and the IC last.* The failure rate for ICs, even if they are surplus or factory "seconds," is quite low. If you still think the IC is at fault, try substitution, or if the IC cannot be easily removed, "piggyback" another IC atop the suspect one so that their pins touch and match exactly.

If a circuit is not functioning at all, the first thing to check is the power supply and the supply voltage and ground connections. A volt-ohmmeter (VOM) or similar "multitester" can quickly show the voltage being supplied and whether it is present at certain points. Make certain that the proper supply voltage is in fact being supplied. Other components, such as resistors, capacitors, and the connecting wires between the various parts of the circuit, should be tested. If all seem okay, the ICs can be checked.

More commonly, the circuit will function but not in the manner desired. In such cases, the components are almost invariably satisfactory, but you have done or are doing something wrong.

The first place to look is at how you have constructed the circuit. Have you observed the precautions given in previous chapters for using certain devices? For example, you may have left an IC input "floating" or held to an incorrect logic state. Or you may have forgotten to use "de-glitching" capacitors with TTL devices, or the leads connecting components in a high-frequency circuit may be too long. If you are using an ac power supply, it may be too noisy—try substituting a battery power supply and see if the problem clears up. Also check that you have bypassed power supply connections to the circuit using capacitors where suggested.

The clock frequency source can be the cause of many problems in digital circuits. A clock generator with a variable frequency is a big help here. Try varying the frequency of the clock signal to see if this has any impact on the operation of the circuit.

If none of the preceding clear up the problem, you will have to *trace* the operation of the circuit until you locate the cause of the trouble. You can begin the trace at either the input or output of a circuit; the choice of where to begin is based on common sense and, often, intuition. For example, suppose that you have a circuit whose outputs drive some LEDs to indicate circuit operation. The obvious place to begin is at the final outputs and then work "backwards" in the logic path until you find the trouble. If you have a device at the "beginning" of a circuit whose outputs drive several other devices, the best place to begin would be with that device and then work forward. *Any* device whose outputs drive several other devices is an immediate suspect, regardless of where it is in the circuit. Try checking the inputs to that device to see if they are correct; if so, check each of its outputs.

There are two essentials in debugging digital circuits. One is a steady, "clean" source of high and low logic signals. The "bounceless" and selectable output switches discussed in Chapters 4 and 5 are ideal for this purpose. The other necessity is a *logic probe.* A logic probe is a device with LEDs that indicate the state of a logic signal applied to it; a low signal does not light the LED, while a high signal does. Many logic probes include another LED which flickers when the input signal is constantly switching states. Better logic probes also include a "pulse stretcher" to indicate very short pulses and a "memory" to hold the state of the LEDs until reset. Logic probes are small devices, often only slightly larger than a pen, which can be held in the hand and can easily reach and touch the pins of an IC. Logic probes are available from electronics parts retailers for a few dollars and are well worth the money.

You can also improvise various logic state indicators using some of the logic gate circuits described in Chapters 4 and 5. These are useful since logic probes are single-input devices and you may wish to test whether or not several different signals are present.

A final point to remember in circuits using resistors and capacitors is that components usually vary significantly from their nominal value. For example, many resistors sold to hobbyists and experimenters may vary 20% above or below their assigned value. Capacitors also vary from their assigned values and their value often changes over time. If a circuit fails to oscillate at a desired frequency, try substituting another part of the same or similar value. (This is especially true in circuits where a pair of capacitors of the same value is used.) Try using potentiometers in place of fixed-value resistors when experimenting with circuits.

Finally, if a problem really has you baffled, try putting the circuit away—out of your sight—for a couple of days. It is amazing how many wiring errors and similar gaffes become obvious once you have forgotten what the circuit is "supposed" to look like!

YOUR NEXT STEPS

This book has been only a sample of the wide variety of IC devices and technologies available. Now it is up to you to keep current with new devices and technologies and to expand your skills into other areas.

TABLE 6-1 Addresses of Major IC Manufacturers

Advanced Micro Devices 901 Thompson Place Sunnyvale, CA 94088	Linear Technology 1630 McCarthy Boulevard Milpitas, CA 95035-7487
Analog Devices Route 1 Industrial Park Norwood, MA 02062	Motorola Semiconductor 5005 East McDowell Road Phoenix, AZ 85008
Burr-Brown International Airport Industrial Park Tucson, AZ 85734	National Semiconductor 2900 Semiconductor Drive Santa Clara, CA 95052
Datel 11 Cabot Boulevard Mansfield, MA 02048	Raytheon Semiconductor 350 Ellis Street Mountain View, CA 94043
Exar 750 Palomar Avenue P.O. Box 62229 Sunnyvale, CA 94088	RCA Solid State Box 3200 Somerville, NJ 08876
Fairchild Semiconductor 464 Ellis Street Mountain View, CA 94042	Signetics 811 East Arques Avenue P.O. Box 409 Sunnyvale, CA 94086
Intel 3065 Bowers Avenue Santa Clara, CA 95051	Teledyne Semiconductor 1300 Terra Bella Avenue Mountain View, CA 94043
Intersil 10710 North Tantau Avenue Cupertino, CA 95014	Texas Instruments Semiconductor Group P.O. Box 401560 Dallas, TX 75240

National Semiconductor

February 1984

DS8910 AM-Only Digital Phase-Locked Loop Frequency Synthesizer

DS8910 AM-Only Digital Phase-Locked Loop Frequency Synthesizer

General Description

The DS8910 is a PLL synthesizer designed specifically for use in AM radios. It contains the reference oscillator, a phase comparator, a charge pump, an operational amplifier, a 15 MHz ECL/I^{2}L dual modulus programmable divider, and a 19-bit shift register/latch for serial data entry. The device is designed to operate with a serial data controller generating the necessary division codes for each frequency, and logic state information for radio function inputs/outputs.

A 3.96 MHz pierce oscillator and divider chain generate a 1.98 MHz external controller clock, a 20 kHz, 10 kHz, 9 kHz, and 1 kHz reference signals, and a 50 Hz time-of-day signal. The oscillator and divider chain are sourced by the V_{CCM} pin thus providing a low power controller clock drive and time-of-day indication when the balance of the PLL is powered down.

The 21-bit serial data stream is transferred between the frequency synthesizer and the controller via a 3-wire bus system comprised of a data line, a clock line, and an enable line.

The first 2 bits in the serial data stream address the synthesizer thus permitting other devices such as display drivers to share the same bus. The next 14 bits are used for the PLL (N + 1) divide code. The 15th bit is a "don't care" bit and serves no function. The 16th and 17th bits are used to select one of the 4 reference frequencies. The 18th and 19th bits are connected via latches to open collector outputs. These outputs can be used to drive radio functions such as gain, mute, AM, or charge pump current source levels.

The PLL consists of a 14-bit programmable I^{2}L divider, an ECL phase comparator, an ECL dual modulus (p/p + 1) prescaler, a high speed charge pump, and an operational amplifier. The programmable divider divides by (N + 1), N being the number loaded into the shift register. The programmable divider is clocked through a ÷ 7/8 prescaler by the AM input. The AM input will work at frequencies up to 15 MHz. The VCO can be tuned with a frequency resolution of either 1 kHz, 9 kHz, 10 kHz, or 20 kHz. The buffered AM input is self-biased and can be driven directly by the VCO through a capacitor. The ECL phase comparator produces very accurate resolution of the phase difference between the input signal and the reference oscillator. The high speed charge pump consists of a switchable constant current source and sink. The charge pump can be programmed to deliver from 75 μA to 750 μA of constant current by connection of an external resistor from pin $R_{PROGRAM}$ to ground or the open collector bit outputs. Connection of programming resistors to the bit outputs enables the controller to adjust the loop gain for the particular reference frequency selected. The charge pump will source current if the VCO frequency is high and sink current if the VCO frequency is low. The low noise operational amplifier provided has a high impedance JFET input and a large output voltage range. The op amp's negative input is common with the charge pump output and its positive input is internally biased.

Features

- Uses inexpensive 3.96 MHz reference crystal
- Serial data entry for simplified control
- 50 Hz output for time-of-day reference driven from separate low power V_{CCM}
- 2 open collector buffered outputs for controlling various radio functions or loop gain
- AM input has 15 mV (typical) hysteresis
- Programmable charge pump current sources enable adjustment of system loop gain
- Operational amplifier provides high impedance load to charge pump output and a wide voltage range for the VCO input

Connection Diagram

Dual-In-Line Package

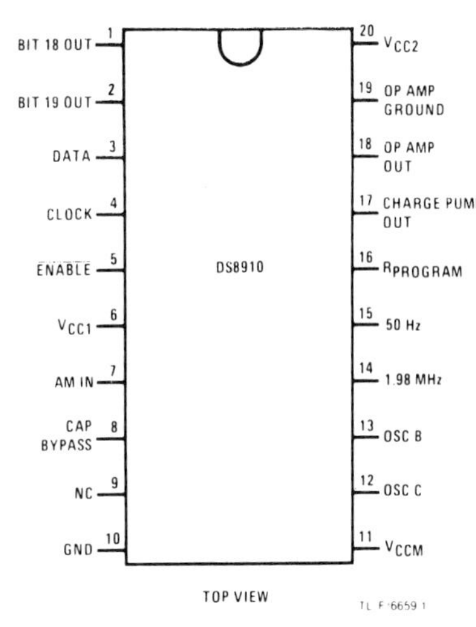

Circle DATA UPDATE No. 103660

Figure 6-6 Typical manufacturer's data sheet for an IC.

Throughout this book, we have mentioned "data sheets" and "applications notes." Figure 6-6 shows a typical manufacturer's data sheet. These give the essential information, such as pin connections, internal diagram, specifications, and so on, for the device in question. The application note assumes that you already have the data sheet, and instead, covers the actual use of the device. (Unfortunately, not all devices have applications notes, although all have data sheets.)

Most manufacturers will send a single data sheet or "ap (applications) note" upon request, although it often helps if you make your request on a business letterhead. Many manufacturers have compiled their data sheets and ap notes together into what are known as "data books." Although useful, these books can be expensive, and a complete set from all manufacturers will cost several hundreds of dollars. Fortunately, compilations of the data sheets and ap notes for the most widely used ICs are available from independent publishers.

Table 6-1 lists the addresses of some of the larger IC manufacturers. A letter to their Literature Distribution Department will get you information about ordering their data books and getting on their mailing lists for future data sheets and applications notes.

AND MOST IMPORTANT . . .

Have fun with IC technology. Experiment. Grow. Enjoy!

index